D0527716
Target
Get back on track
7
AQA GCSE (9–1)
Physics
Darren Forbes, Alex Holmes,
Gillian Nightingale, Helen Sayers
and Jason Welch
Pearson

Published by Pearson Education Limited, 80 Strand, London, WC2R 0RL.

www.pearsonschoolsandfecolleges.co.uk

Typeset and illustrated by QBS Learning
Produced by QBS Learning

The rights of Darren Forbes, Alex Holmes, Gillian Nightingale, Helen Sayers and Jason Welch to be identified as authors of this work have been asserted by them in accordance with the Copyright, Designs and Patents Act 1988.

First published 2018

21 20 19 18
10 9 8 7 6 5 4 3 2 1

British Library Cataloguing in Publication Data
A catalogue record for this book is available from the British Library

ISBN 978 1 292 24578 2

Printed in Slovakia by Neografia

Note from the publisher
Pearson has robust editorial processes, including answer and fact checks, to ensure the accuracy of the content in this publication, and every effort is made to ensure this publication is free of errors. We are, however, only human, and occasionally errors do occur. Pearson is not liable for any misunderstandings that arise as a result of errors in this publication, but it is our priority to ensure that the content is accurate. If you spot an error, please do contact us at resourcescorrections@pearson.com so we can make sure it is corrected.

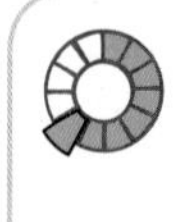

This workbook has been developed using the Pearson Progression Map and Scale for Science.

To find out more about the Progression Scale for Science and to see how it relates to indicative GCSE (9–1) grades go to www.pearsonschools.co.uk/ProgressionServices

Helping you to formulate grade predictions, apply interventions and track progress.

Any reference to indicative grades in the Pearson Target Workbooks and Pearson Progression Services is not to be used as an accurate indicator of how a student will be awarded a grade for their GCSE exams.

You have told us that mapping the Steps from the Pearson Progression Maps to indicative grades will make it simpler for you to accumulate the evidence to formulate your own grade predictions, apply any interventions and track student progress. We're really excited about this work and its potential for helping teachers and students. It is, however, important to understand that this mapping is for guidance only to support teachers' own predictions of progress and is not an accurate predictor of grades.

Our Pearson Progression Scale is criterion referenced. If a student can perform a task or demonstrate a skill, we say they are working at a certain Step according to the criteria. Teachers can mark assessments and issue results with reference to these criteria which do not depend on the wider cohort in any given year. For GCSE exams however, all Awarding Organisations set the grade boundaries with reference to the strength of the cohort in any given year. For more information about how this works please visit: https://www.gov.uk/government/news/setting-standards-for-new-gcses-in-2017

Contents

Get started **AO1, AO2, AO3**

1 Rotational forces

This unit will help you to learn more about turning forces, and how to use moments to calculate the force needed to balance some levers.

In the exam, you will be asked to answer questions such as the one below.

Exam-style question

1 **Figure 1** shows a see-saw.

1.5 m *d*
350 N 500 N
not to scale pivot

Figure 1

1.1 Calculate the moment of the 350 N force about the pivot. **(4 marks)**

Give the unit.

Moment = ... Unit: ....

1.2 Calculate the distance, *d*, in metres, that the 500 N force needs to be from the pivot to balance this see-saw. **(3 marks)**

Distance = ... m

You will already have done some work on rotational forces. Before starting the **skills boosts**, rate your confidence in these areas. Colour in the bars.

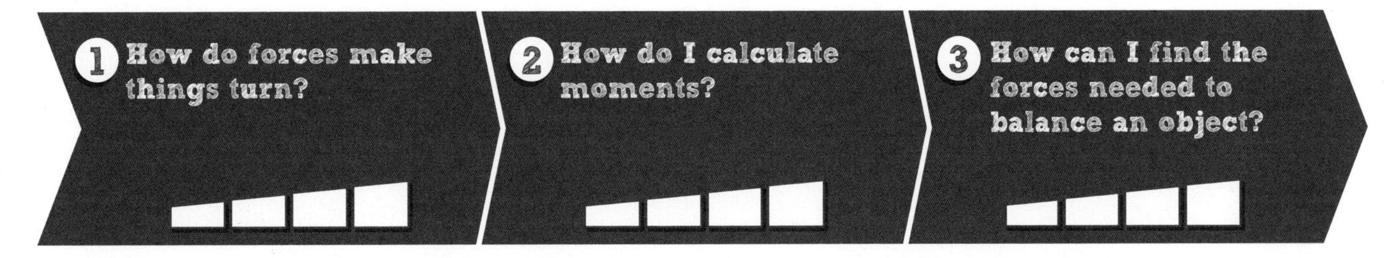

Get started

It is important to use the correct units when calculating rotational forces. There is also specific vocabulary to learn.

1 Draw a line to match each word with its meaning. One has been done for you.

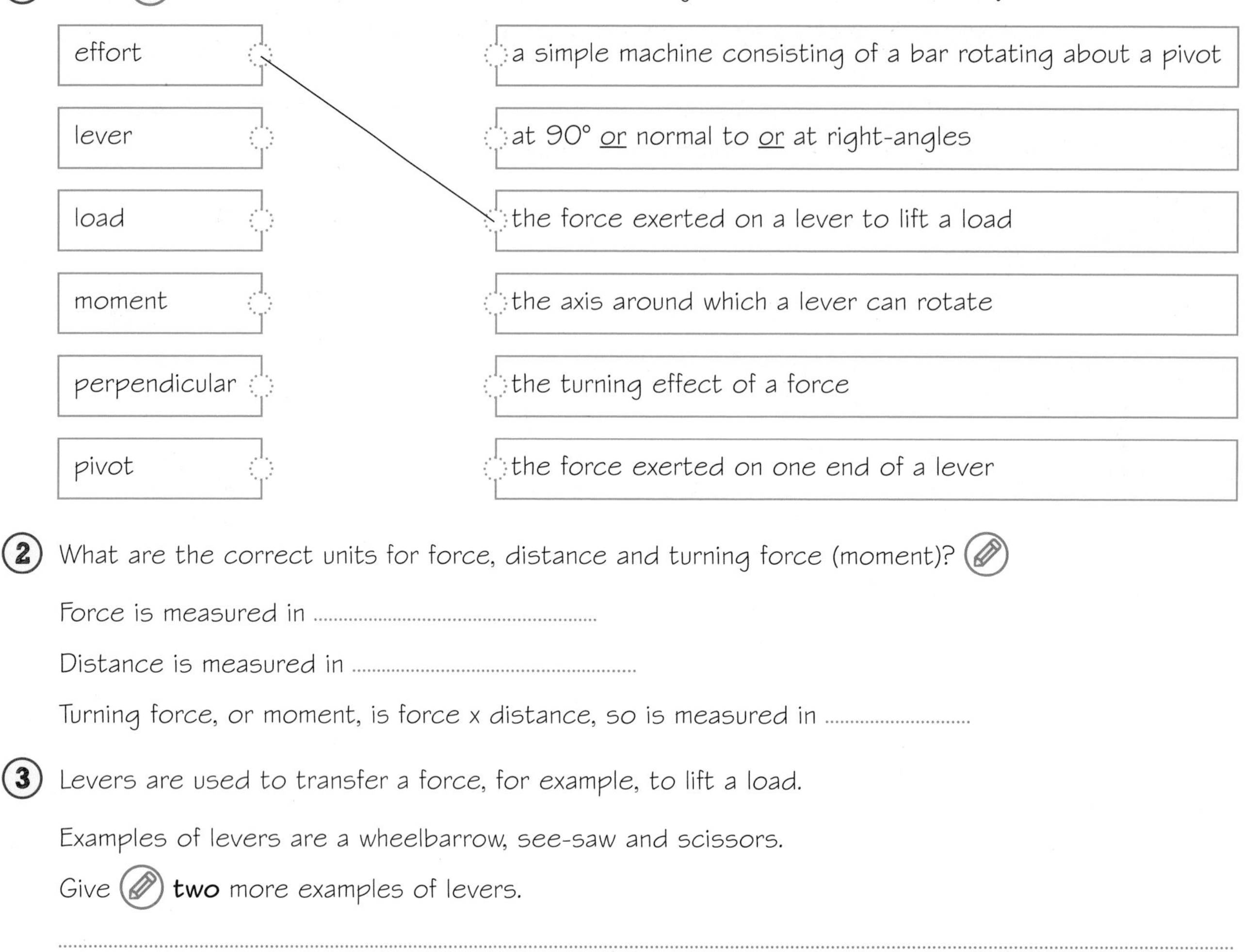

2 What are the correct units for force, distance and turning force (moment)?

Force is measured in ....

Distance is measured in ....

Turning force, or moment, is force x distance, so is measured in ....

3 Levers are used to transfer a force, for example, to lift a load.

Examples of levers are a wheelbarrow, see-saw and scissors.

Give **two** more examples of levers.

....

4 Gears consist of toothed wheels fixed to shafts. The teeth interlock with each other. When one wheel turns around the shaft, it causes the other wheel to turn as well. The turning force is transmitted from one wheel to the other. The motion of the wheels will vary in speed and direction (clockwise or anticlockwise).

Two examples of gears are shown below.

Cross out ~~cat~~ the incorrect words to complete the sentences.

a When the large wheels rotate anticlockwise, the small wheels rotate **clockwise / anticlockwise**.

b The movement of the wheels is perpendicular to each other for gear **type 1 / type 2**.

c For type 1, the **small / big** gear will do the most turns.

d For type 2, the **small / big** gear will do the most turns.

Skills boost

How do forces make things turn?

Forces can make things turn when there is a pivot to rotate about. The rotational effects of forces can be transmitted through simple machines such as levers and gears.

The first diagram shows three different ways that turning forces can be transmitted with levers. There is no turning effect if the effort force acts straight through the pivot.

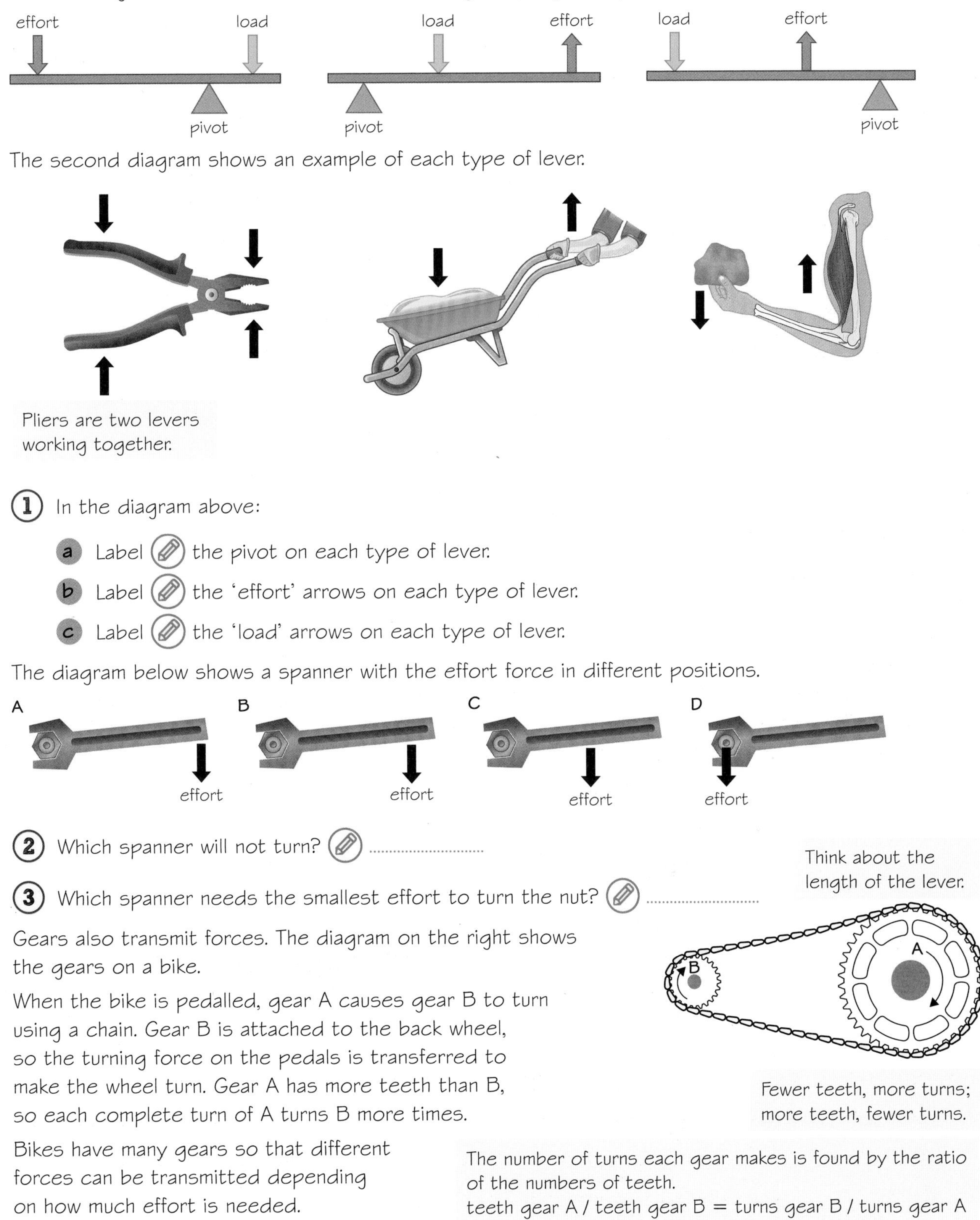

The second diagram shows an example of each type of lever.

Pliers are two levers working together.

1 In the diagram above:

- **a** Label the pivot on each type of lever.
- **b** Label the 'effort' arrows on each type of lever.
- **c** Label the 'load' arrows on each type of lever.

The diagram below shows a spanner with the effort force in different positions.

2 Which spanner will not turn? ....

3 Which spanner needs the smallest effort to turn the nut? ....

Think about the length of the lever.

Gears also transmit forces. The diagram on the right shows the gears on a bike.

When the bike is pedalled, gear A causes gear B to turn using a chain. Gear B is attached to the back wheel, so the turning force on the pedals is transferred to make the wheel turn. Gear A has more teeth than B, so each complete turn of A turns B more times.

Fewer teeth, more turns; more teeth, fewer turns.

Bikes have many gears so that different forces can be transmitted depending on how much effort is needed.

The number of turns each gear makes is found by the ratio of the numbers of teeth.
teeth gear A / teeth gear B = turns gear B / turns gear A

2 How do I calculate moments?

The turning effect of a force is called the moment of the force.

You need to learn and be able to use this equation to calculate the moment of a force:

moment of a force (newton-metre, Nm) = force (newton, N) × distance (metre, m)

The distance, *d*, is the perpendicular distance from the pivot to the line of action of the force.

1. A person pushes a door with a force of 20 N.

Don't confuse moment with momentum, which is the mass of an object multiplied by its velocity.

a. Calculate the moment when the door is pushed 40 cm from the hinge.

To convert cm to m divide by 100.

i Convert the distance into m as the standard unit. 40 cm = ... m

ii Moment (Nm) = ... N × .... m = ... Nm

Substitute the values into the equation.

b. Calculate the moment when the door is pushed 80 cm from the hinge.

Moment (Nm) = ... N × .... m = ... Nm

Check the units.

c. The distance of the force from the hinge has doubled. What is the relationship between the moment and the distance for the same force? Complete the sentence.

Whatever happens to one side of the equation must happen on the other.

When the distance doubles, the moment ....

2. A painter uses a screwdriver to open a tin of paint.
An effort force of 15 N produces a moment of 1.5 Nm.
Calculate the length of the screwdriver.

length of screwdriver
15 N

a. On the diagram, label the pivot, P.

b. Highlight the values you are given in the question.

c. The equation to calculate moment is
moment (Nm) = force (N) × distance (m)
Circle the value in the equation that you are asked to calculate.

d. i Substitute the values into the equation.

The distance is going to be the length of the screwdriver.

.... = ... × distance (m)

ii Solve the equation.

Divide both sides by the force (N).

$\frac{\square}{\square}$ = distance (m)

iii Calculate the answer.

.... m = distance (m)

length of screwdriver = ... m

Skills boost

How can I find the forces needed to balance an object?

When rotational forces are in equilibrium, an object will be balanced. We use the principle of moments to find out whether an object will turn one way, or the other, or be balanced.

You need to recall and use the principle of moments for rotational forces:

> If an object is balanced, the total clockwise moment about a pivot equals the total anticlockwise moment about that pivot.
>
> the sum of clockwise ↻ moments = the sum of anticlockwise ↺ moments

How to use the principle of moments:

1. Calculate the sum of the moment(s) in the direction that the force <u>and</u> distance are known.
2. If the object is balanced, the sum of the other moments in the opposite direction will be the same.
3. Use the moment (Nm) to calculate the distance (m) if the force (N) is known, or the force (N) if the distance (m) is known.

1 The diagram shows a crane. The arm of the crane is balanced when it lifts different loads.

3 m
6 m
10 000 N
counterweight
loading platform
F

a Circle Ⓐ the pivot on the diagram.

b Circle Ⓐ the correct arrow to show which way this crane will tip if there is no loading platform. ↻ / ↺

c Calculate the clockwise moment.

moment = force × distance

clockwise moment = ... N × .... m = ... Nm

d Calculate the load force, *F*, that can be lifted by this crane when the loading platform is 6 m from the tower.

The crane will be balanced when the anticlockwise moment is the same as the clockwise moment.

anticlockwise moment = clockwise moment = ... Nm

anticlockwise moment = *F* × .... m = ... Nm

Rearranged equation:

$$\text{force} = \frac{\text{moment}}{\text{distance}} = \frac{\square}{\square}$$

= .... N

e On paper, calculate the distance the loading platform should be from the tower to safely lift a force of 2000 N.

Follow the steps in **d** to help you.

Sample response

Use the sample student response to improve the way you answer questions about rotational forces.

Exam-style question

1 **Figure 1** shows a crowbar.

The force is exerted 50 cm from the pivot.

1.1 Calculate the force that needs to be exerted to produce a moment of 80 Nm at the pivot. **(3 marks)**

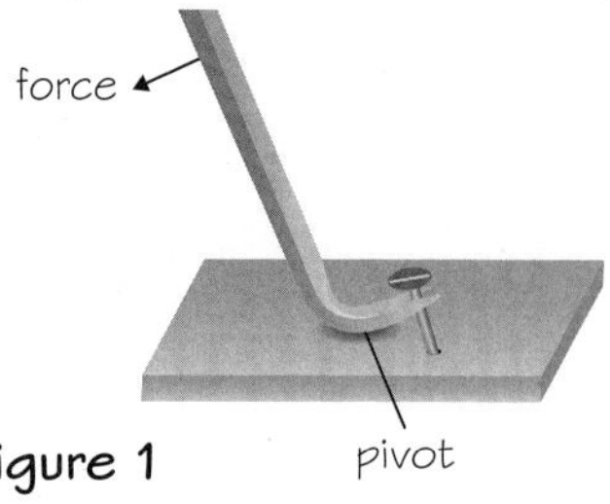

Figure 1

1 Highlight the values given in the question, including the units.

2 **a** What are the units used for moments? ....

b What are the units used for force? ....

c What are the units used for distance when calculating moments? ....

Look at this student's answer to **1.1**.

80 = force × 50 *force =* $\frac{80}{50}$ *= 1.6 N*

3 Circle the error the student has made.

4 Now answer **1.1** yourself. Include the equation you will use.

Exam-style question

1.2 Describe how a larger moment could be produced. **(2 marks)**

Here is a sample student answer to **1.2**.

Make the force bigger.

5 Highlight the quantity in the student answer.

6 Which other quantity should the student have included in their answer? ....

....

7 Complete these sentences to describe how a larger moment could be produced.

Use a .... force for the same (or a larger) distance.

Use the .... force but increase the .... the force is applied at.

8 **a** How many marks would you give this answer? ....

b Why would you give this mark?

....

....

Increase both the force and the distance the force is applied at.

Your turn!

It is now time to use what you have learned to answer the exam-style question from page 1. Remember to read the question thoroughly, looking for information that might help you. Make good use of your knowledge from other areas of physics.

Read the exam-style question and answer it using the guided steps below.

Exam-style question

1 **Figure 1** shows a see-saw.

1.1 Calculate the moment of the 350 N force about the pivot. **(4 marks)**

Give the unit.

Moment = ... Unit: ....

Figure 1

1.2 Calculate the distance, *d*, in metres, that the 500 N force needs to be from the pivot to balance this see-saw. **(3 marks)**

Distance = ... m

1.1 i What is the equation you need to use?

....

ii Highlight the values you need in the exam-style question.

iii Substitute the values into the equation.

iv Calculate your answer and include the unit for moments.

Moment = ... Unit: ....

1.2 i What is the value of the anticlockwise moment? ....

ii What is the relationship between the clockwise and anticlockwise moments when the load is balanced?

....

iii Highlight the value(s) you need in the exam-style question above.

iv Rearrange the equation to make distance the subject.

v Substitute the values into the equation.

vi Calculate the answer. The unit has been given for you.

Distance = ... m

Need more practice?

Exam questions may ask about different parts of a topic, or parts of more than one topic. Questions about rotational forces could occur as:

- questions about that topic only, including calculations
- part of a question involving other forces
- part of a question about an experiment or investigation.

Have a go at this exam-style question.

Exam-style question

1 **Figure 1** shows a see-saw with two people sitting on it.

1.1 Calculate the anticlockwise moment about the pivot. **(3 marks)**

Anticlockwise moment = ... Nm

2 m 2 m

650 N pivot 500 N

Figure 1

Write down the equation you need to use.

Check you are using the correct units.

1.2 Another person weighing 200 N gets on the right-hand side of the see-saw.

Calculate the distance from the pivot they should sit to balance this see-saw. **(4 marks)**

sum of anticlockwise moments = sum of clockwise moments (when something is balanced)

Distance = ... m

Boost your grade

Practise making up your own moment calculations using different types of levers and make sure you can rearrange the equations.

How confident do you feel about each of these **skills**? Colour in the bars.

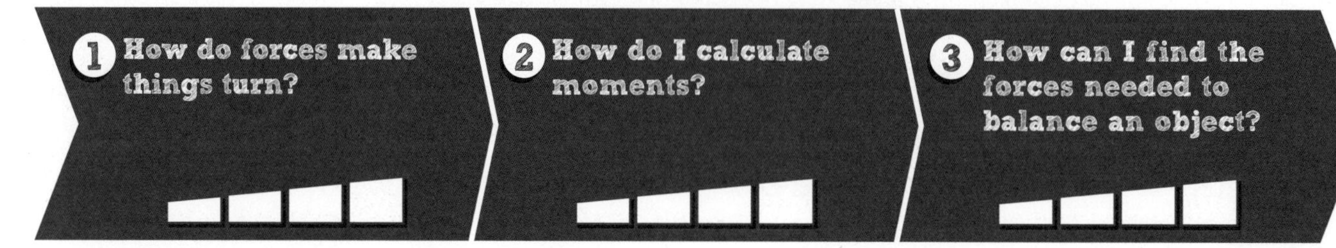

AO2

2 Lenses

This unit will help you to understand how light rays help to explain refraction and to show how lenses work.

In the exam, you will be asked to answer questions such as the one below.

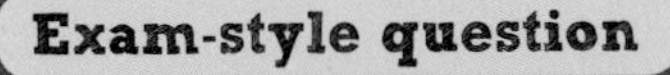

1 **Figure 1** shows a convex lens used to view a small object.

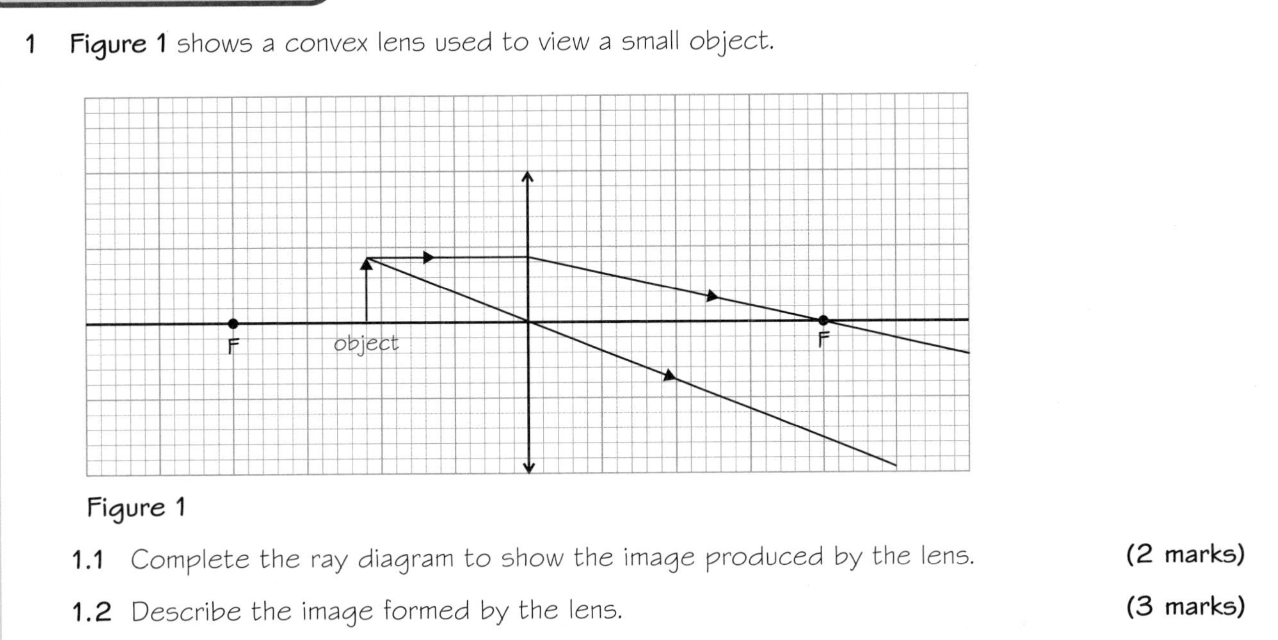

Figure 1

1.1 Complete the ray diagram to show the image produced by the lens. **(2 marks)**

1.2 Describe the image formed by the lens. **(3 marks)**

You will already have done some work on light rays and refraction. Before starting the **skills boosts**, rate your confidence in using light rays to solve problems about lenses. Colour in the bars.

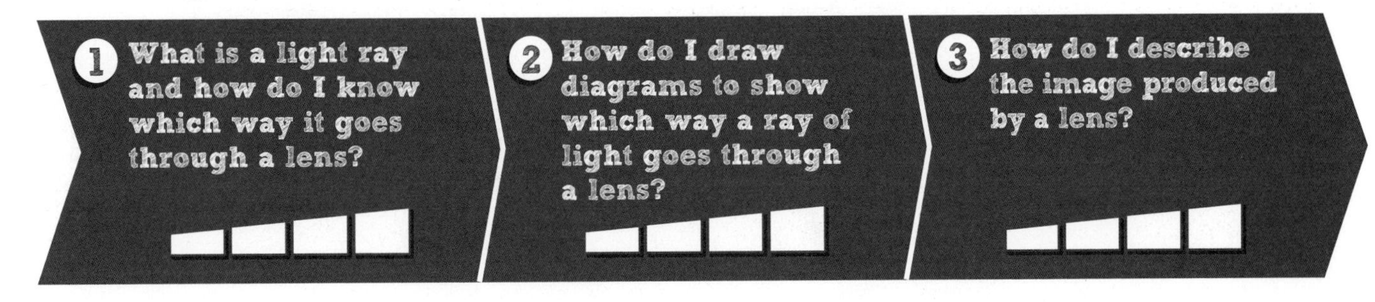

To understand how images are formed by lenses, you need to revise some basic ideas about light rays and refraction.

1 The diagram shows a light ray travelling from a light source (lamp) into a detector (a person's eye).

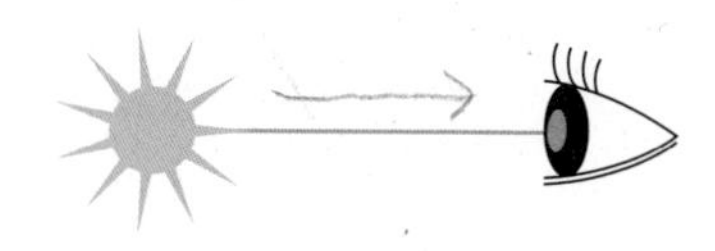

Draw an arrow on the light ray.

2 Complete the sentence using words from the box to explain what the diagram shows.

detector	energy	light	source	waves

Light travels from a ......Source...... to a ......detector......

A light ray shows the direction of transfer of ......energy......

When light meets a new material, some energy is transmitted, some is reflected and some is absorbed.

3 Label the three arrows in the diagram with the words reflect, absorb and transmit.

Light travels at different speeds in different materials. Because of this, transmitted light can change direction when it enters a new material – this is called refraction.

4 **a** Complete the diagram showing refraction of a light ray entering a transparent plastic block.

b Add the letters *i* and *r* to the diagram to show the angle of incidence and angle of refraction.

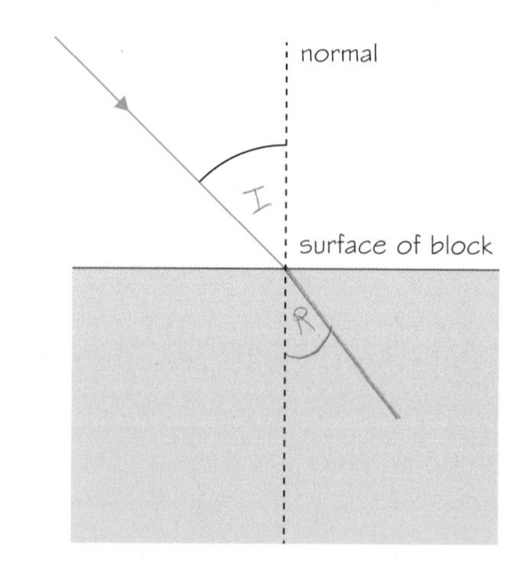

The normal is a construction line drawn at right angles to the surface where a light ray meets the surface.

Light **slows down** when moving from air into glass/plastic.
Light **speeds up** when moving from glass/plastic into air.

5 Circle the correct words in bold to describe how this affects light rays.

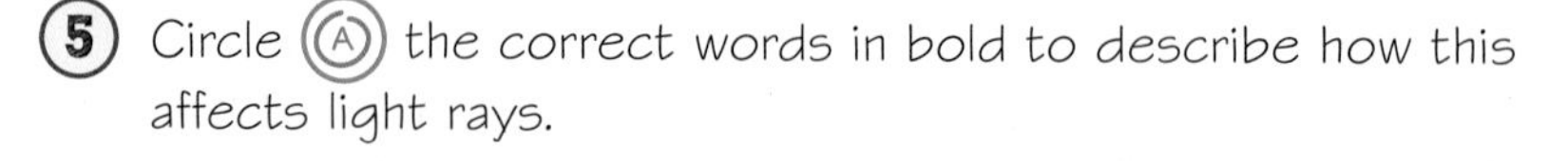

The angle of refraction is **the same as** / **bigger than** / **smaller than** the angle of incidence when light passes from air to glass.

6 The plastic block is now turned on its side. Light will bend towards the normal when it goes into the plastic block. Light will bend away from the normal when it moves into the air.

Complete the diagram to show the direction the light ray leaves the block on the right.

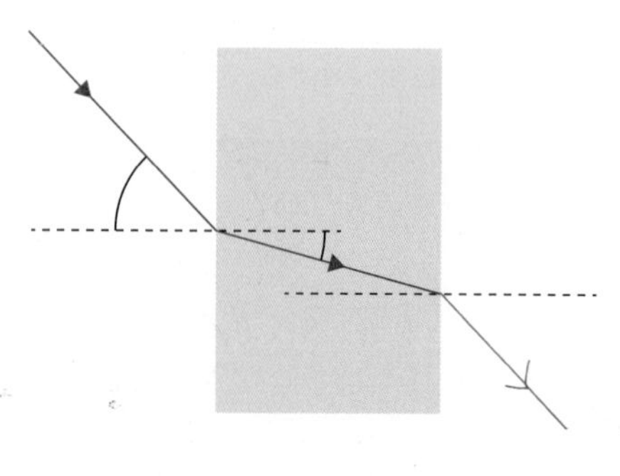

1 What is a light ray and how do I know which way it goes through a lens?

Rays are lines drawn to show the direction in which energy is travelling. Refraction can make light rays **converge** (come together) or **diverge** (spread apart).

Look at the picture of the person's face. Light from a light source is reflected off the person in all directions. (If it wasn't, then you wouldn't be able to see them!)

1 Add some rays with arrows to the diagram to show light reflected off the person.

When working out what happens to light, there's no need to track all the rays. You only need to draw one or two rays. This can be confusing if you think the drawn ray is the only one – it isn't. You just need to remember that all the other rays are doing something similar.

2 Complete the green rays in the diagram to show how they do the same as the blue ray.

Include arrows on your light rays.

The angle of incidence equals the angle of reflection.

Look at the diagram on the right. A light ray is refracted through a triangular-shaped piece of plastic. Notice that there is a normal line drawn for the light ray entering the plastic and a different normal when the ray leaves the plastic.

normal

normal

3 a Complete the sentence.

The normal line is always at right-angles to Surface.

Look back at page 10.

b Complete the path of the green ray.

The two light rays should come together – they converge.

The completed diagram will be symmetrical; the black line in the centre is the line of symmetry.

Swapping the triangle-shaped plastic pieces around makes the top and bottom rays change places.

4 Complete the diagram below by completing the paths of the light rays.

Add two normal lines for each ray.

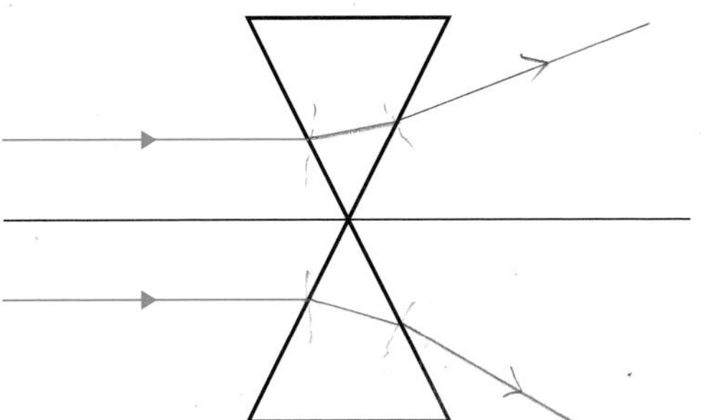

If a transparent material is shaped like one of the diagrams above it is called a **lens**.

Skills boost

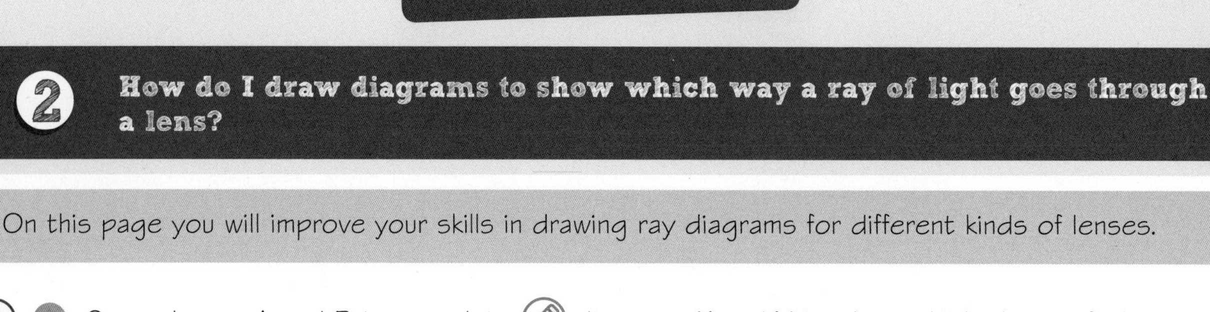

2 How do I draw diagrams to show which way a ray of light goes through a lens?

On this page you will improve your skills in drawing ray diagrams for different kinds of lenses.

1 **a** Copy shapes A and B to complete diagrams **X** and **Y** to show which shape of glass would affect the light rays as shown.

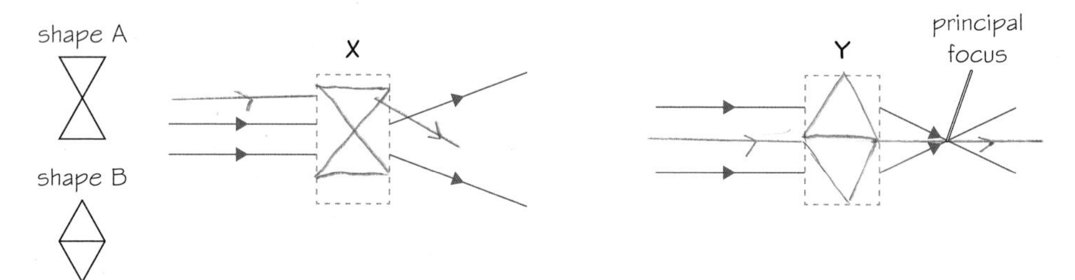

b The rays shown above are not the only rays arriving at the lenses. Add two more light rays parallel to the others on each diagram and draw their paths through the lenses.

Don't forget to add arrows.

Shape A (thin in the middle) is called **concave**. Shape B (thin at the ends) is called **convex**. Lenses used in telescopes and projectors do not use triangle shapes – they have curved sides (these are easier to make).

2 Draw lines to link the shape to its name and symbol.

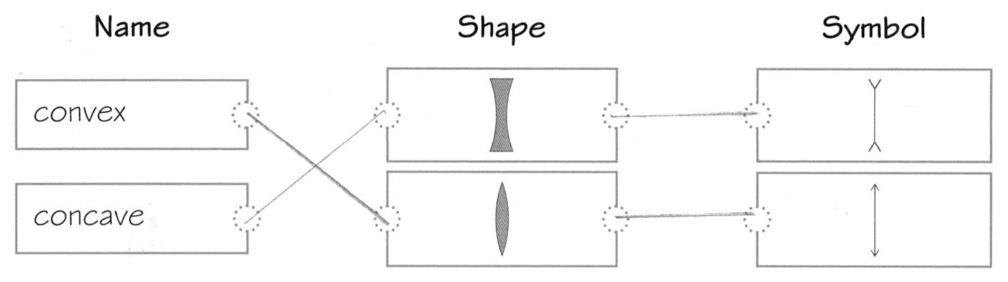

3 **a** Add the correct symbol for a convex lens to the right of these parallel rays.

b Continue the rays through the lens to show where they go and mark the principal focus with an **F**.

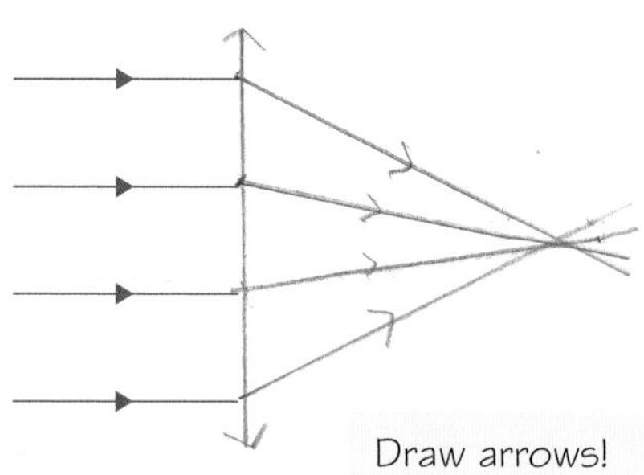

Draw arrows!

Here is a diagram for a concave (diverging) lens. The parallel light rays do not meet on the right.

To find the principal focus of a concave lens, you need to 'dot back' the rays by placing a ruler along a ray and continuing the ray to the left with a dotted line.

4 **a** Ray **A** has been done for you. Place a ruler along ray **B** and continue the ray to the left with a dotted line.

b Mark the principal focus with a letter **F**.

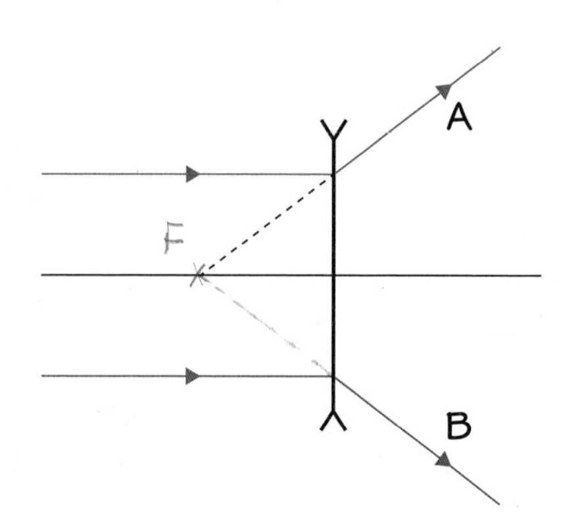

3 How do I describe the image produced by a lens?

When you look through a lens you see a distorted image: things might look a different size or might appear upside-down. To describe the image, you need to decide:

- Is the image bigger than the object (magnified) or smaller (diminished)?
- Is the image the right way up (upright) or upside-down (inverted)?
- Is the image real or virtual?

You can use ray diagrams to work out what the image will look like.

In this diagram, the vertical arrow on the left represents the stick figure.

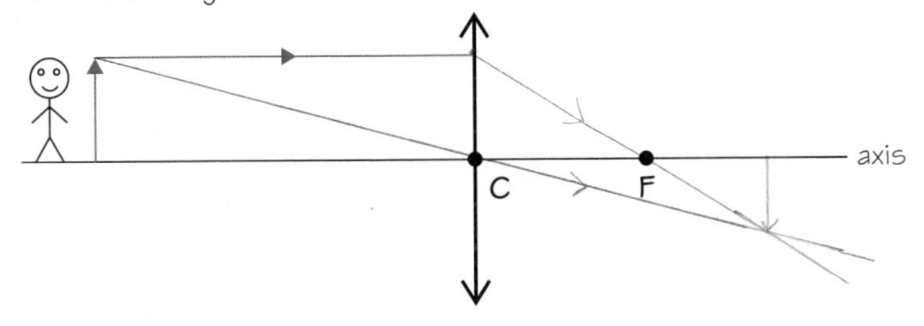

To complete a ray diagram to find an image:

1. Draw a ray from the top of the object, parallel to the axis.
2. Continue it on the other side so it goes through the focus; don't forget the arrows.
3. Draw another ray from the top of the object, straight through the middle of the lens.
4. Mark the place where the two rays cross.
5. Join this point vertically back to the axis.

1 **a** Continue the light ray as it passes through the lens.

Is the lens convex or concave?

b Draw another light ray from the top of the vertical arrowhead straight through the centre of the lens (C).

Always draw rays from the tip of the arrowhead (the top of the object).

c Mark the point, P, where the two rays meet on the right of the diagram. Join this point to the axis with a vertical arrow, making sure the arrowhead points away from the axis. This arrow represents the size, position and orientation of the image.

Don't forget arrows!

An image formed by light rays that come together is called a **real image**.

d Circle the correct words in bold to complete the sentence.

In the diagram above, the image is **inverted** / **upright** and **magnified** / **diminished**.

Sometimes the two rays do not cross after passing through the lens. If this happens, you need to 'dot them back' to find where they seem to meet. When you have to 'dot back' the rays to find the image, the image is **virtual**.

2 **a** Complete the ray marked **A** by dotting it back to the axis.

b Mark where the two rays meet. Join this point to the axis with a vertical arrow, making sure the arrowhead points away from the axis.

A
F
F
axis
object

c Circle the correct words in bold to complete the sentence.

The image is **upright** / **inverted** and **diminished** / **magnified**.

Sample response

To answer a question involving a ray diagram, you need to:
- draw two rays from the object to the lens, one parallel and one straight through the centre
- find where they meet (you may have to 'dot back' one or both of them)
- draw the image and describe it.

Look at this exam-style question and the answers given by students.

Exam-style question

1 **Figure 1** shows an object viewed through a lens.

1.1 Name the type of lens shown in the diagram. Convex **(1 mark)**

1.2 Continue the light rays through the lens to locate the image. **(2 marks)**

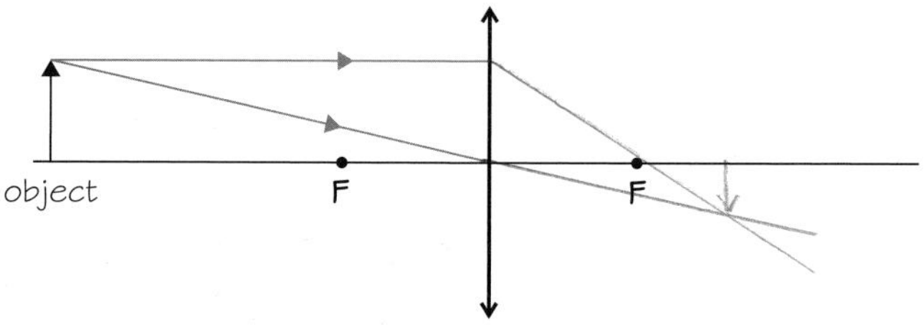

object F F

Figure 1

Here is part of one student's response.

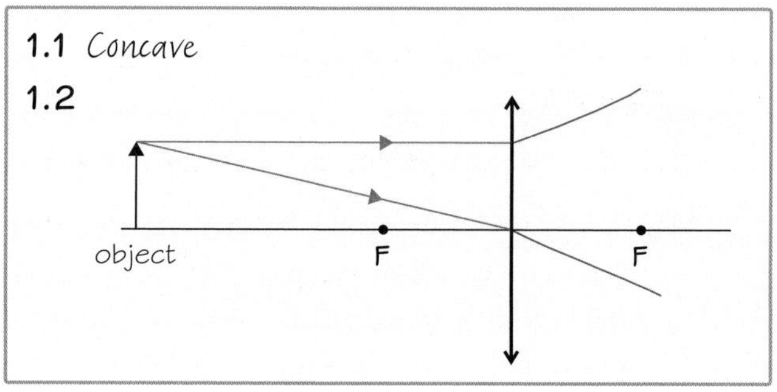

1.1 Concave

1.2

object F F

① **a** Has the student drawn the rays in the correct directions for a concave lens? Yes / No

b What type of lens does the symbol ↕ stand for? Convex

c What has the student forgotten to add to their light rays? Arrows

Here is another student's response to **1.2**.

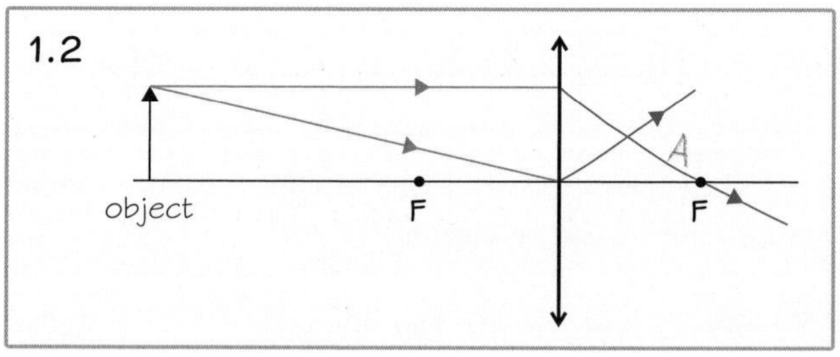

1.2

A

object F F

② This student has drawn one ray in the correct direction. Label this ray A.

Below is a third student's response.

③ This student has correctly identified the lens and drawn the rays. Explain what mistake the student has made in drawing the image.

He needs to draw the image arrow down from the axis so that it points towards the point where both rays intercept eachother. The arrow should also be upside down.

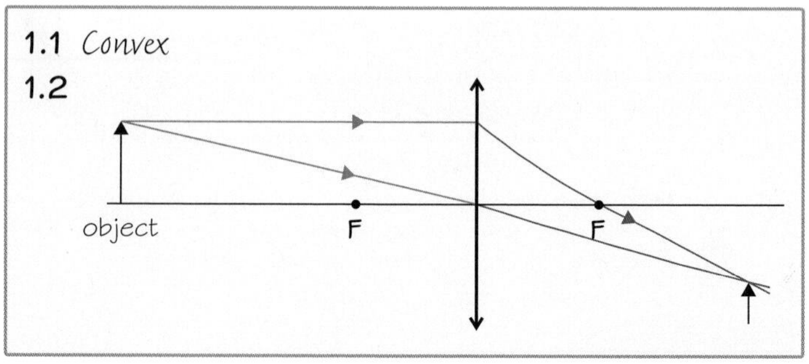

1.1 Convex

1.2

object F F

Where is the axis?

Your turn!

It is now time to use what you have learned to answer the exam-style question from page 9. Remember to read the question thoroughly, looking for information that may help you. Make good use of your knowledge from other areas of physics.

Exam-style question

1 **Figure 1** shows a convex lens used to view a small object.

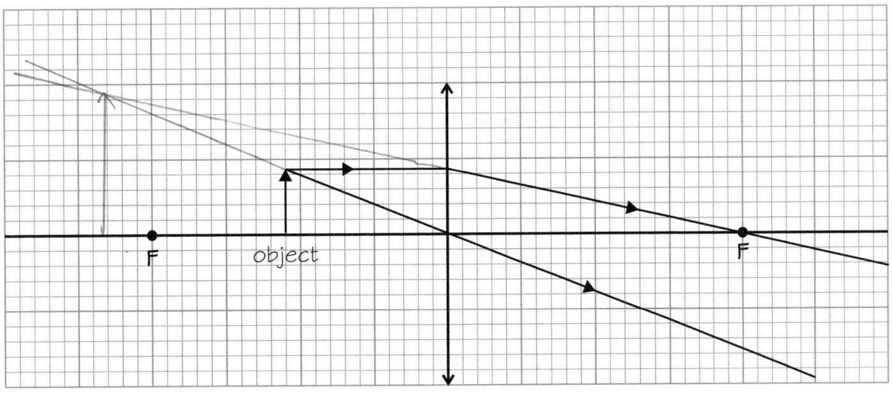

Figure 1

1.1 Complete the ray diagram to show the image produced by the lens. **(2 marks)**

1.2 Describe the image formed by the lens. **(3 marks)**

Magnified + Upright

1. Which kind of lens is shown?

Convex

2. In **Figure 1**, the two rays are drawn for you.

   a Do the rays meet on the right?

   No

   b What do you do when the rays do not meet?

   "Pot them back"

You must show the image on **Figure 1** to get full credit.

3. a Complete the ray diagram to find the image.

   b Write the words that describe the image.

1.2 is worth 3 marks so you need to use three words.

1 Upright

2 Magnified

3 Virtual.

Get started

1 The diagram shows the electromagnetic spectrum.

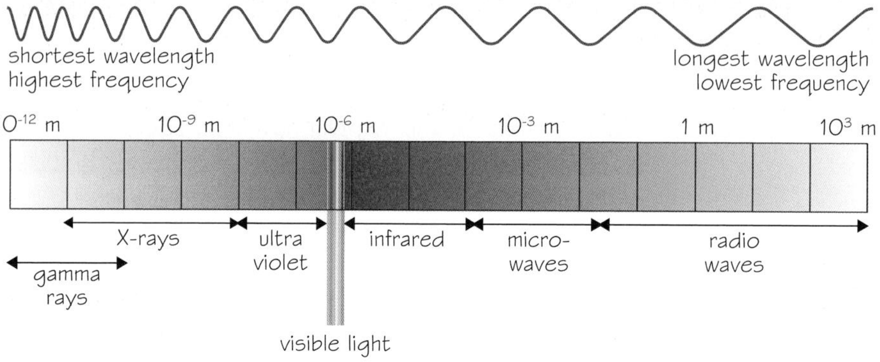

shortest wavelength
highest frequency

longest wavelength
lowest frequency

10^{-12} m | 10^{-9} m | 10^{-6} m | 10^{-3} m | 1 m | 10^{3} m

gamma rays | X-rays | ultra violet | visible light | infrared | micro-waves | radio waves

Visible light contains these seven colours: blue, green, indigo, orange, red, violet, yellow.

a Write the seven colours in order of increasing wavelength.

................................, ..., ..., ..., ...,

................................, ...

b Cross out ~~cat~~ the incorrect words from each pair in **bold** so the sentence is correct.

If light is red-shifted, it moves towards the red end of the spectrum, so the wavelength **increases / decreases** and the frequency **increases / decreases**.

Look at the diagram to help you.

2 If objects are moving at different speeds, or in different directions, or both, then we say they are moving relative to each other.

Tick the **three** statements that show relative movement.

A Two trains travelling along parallel tracks at 45 m/s ☐

B Sound waves from a car radio driving away from you ☐

C Two bicycles racing around a track at 13 m/s and 14 m/s ☐

D Two cars travelling at 27 m/s on a motorway, one 20 m behind the other ☐

E Sound waves from speakers in your ears when you are walking ☐

F Two marbles rolling at the same speed in different directions ☐

3 Draw a line to match each term with its correct meaning.

Term	Meaning
Evidence	A credible conclusion, supported by evidence, can be made.
Peer review	A hypothesis intended to explain something.
Repeatable	The available facts or information confirming that a belief or idea is valid.
Reproducible	The same results are obtained from another person or technique.
Theory	The same results are obtained when the experiment is repeated by the same person.
Valid	A process which checks the work of colleagues.

1 How do I describe red-shift?

If a wave source is moving relative to an observer, there will be a change in the observed frequency and wavelength. Red-shift is a way to describe what happens to waves from a source. This information can be used to provide evidence for different scientific ideas and theories.

You can use other types of waves to help describe the red-shift of light. For example, when a police car goes past, its siren is high pitched as it comes towards you, and becomes low pitched as it moves away. The frequency and wavelength of the sound wave changes. This is called the Doppler effect.

waves closer together

waves further apart

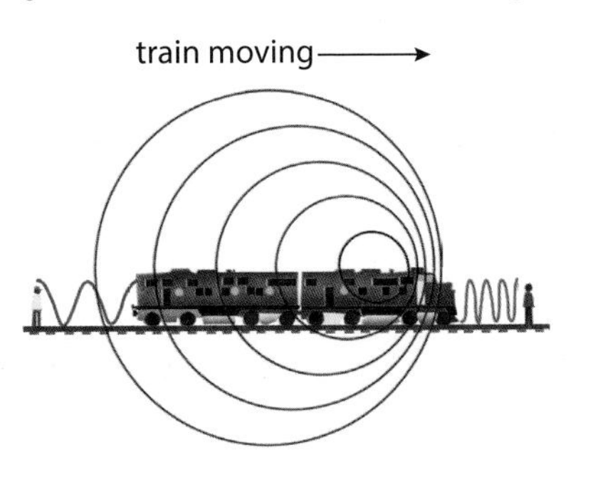

The pitch of the sound you hear will depend on whether the train is moving away from or towards you.

Include the words **wavelength** and **frequency** in your answer.

The wave speed remains the same.

1 Describe the difference in the waves in front of and behind the duck in the diagram above. The duck moves at a steady speed.

....

....

....

2 Cross out ~~cat~~ the incorrect words from each pair in **bold** to complete these sentences.

When a source moves towards an observer, the observed wavelength **decreases / increases** and the frequency **decreases / increases**. When a source moves **towards / away from** an observer, the observed wavelength increases and the frequency **decreases / increases**.

Dark lines are seen in the spectrum of light from different stars. Light from different stars has the same pattern of lines, but the lines are shifted along the scale depending on which way the source moves relative to the observer. Just as the frequency and wavelength of sound or water waves change for an object moving relative to an observer, the same thing happens with light.

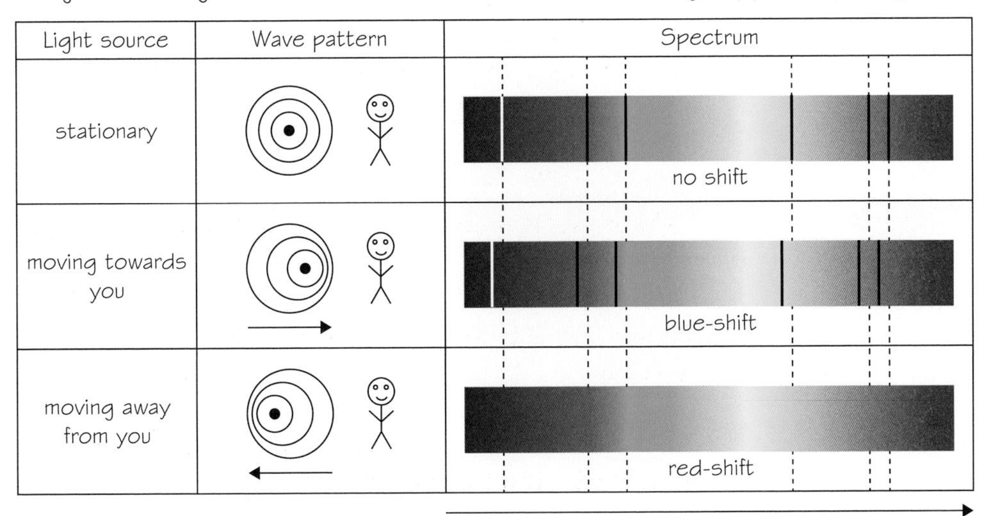

Light source	Wave pattern	Spectrum
stationary		no shift
moving towards you		blue-shift
moving away from you		red-shift

increasing wavelength/decreasing frequency

Blue-shift shows that light moves towards the blue side for sources moving towards the observer. Red-shift is the opposite effect.

3 Draw lines on the red-shift spectrum to show how you predict it will look.

Skills boost

2 How do I explain how red-shift provides evidence for an expanding universe?

The red-shift of light from galaxies provides evidence that the universe is expanding. Scientists use observations to develop theories. Some of the theories scientists have developed about the origin of the universe suggest that it is expanding. The Big Bang theory is one of these theories and is supported by the discovery of red-shift.

1 The greater the red-shift, the faster the galaxy is moving away from us.

a Cross out (~~cat~~) the incorrect words from each pair in **bold** in the table.

Evidence	Interpretation of the evidence
Light / **sound** coming to us from most other galaxies is red-shifted.	Those galaxies are moving **away from** / **towards** us.
The further away a galaxy is, the **more** / **less** the light is red-shifted.	The further away a galaxy is, the **faster** / **slower** it is moving.

b Complete the sentence to summarise this evidence.

The most likely explanation is that the whole universe is getting ....

The evidence of red-shift has been used by scientists to develop theories about the origin of the universe. The fact that a galaxy's speed changes with distance is evidence that the universe is expanding.

The table below summarises information about two respected theories.

Big Bang theory	Steady State theory
First suggested in the 1920s.	First suggested in 1948.
Around 13.5 billion years ago, the universe began from a very small region that was extremely hot and dense. A massive explosion from this single point released huge amounts of radiation. The universe continues to expand from this point.	The universe has always existed.
Gravity caused the matter to come together and form everything in the universe. No new matter is created.	New matter is continuously created within the universe as it expands.

2 Which theory was suggested first? ....

3 Which theory suggests that the universe has always existed? ....

4 Give one similarity between the two theories.

....

....

5 Circle (A) the correct words to show which theories are supported by evidence from red-shift.

Big Bang / Steady State / both

6 Which theory suggests that radiation was released? ....

3 How do I explain why the Big Bang theory is the currently accepted model for the origin of the universe?

Scientists can have different explanations for the same observation. They decide between different models and theories based on the amount of evidence that supports each one. The Big Bang theory is the currently accepted model for the origin of the universe because only the Big Bang theory can explain evidence that scientists have found. This makes it more valid than, for example, the Steady State theory.

1. What evidence supports the idea that the universe is expanding? ....

2. In 1964, cosmic microwave background radiation (CMBR) was detected coming from the whole of space in all directions. Only the Big Bang theory predicts that huge amounts of radiation were released at the start of the universe.

   a. Which theory does the discovery of CMBR support? ....

   b. Which theory is not supported by the discovery of CMBR? ....

3. Choose words from the box to complete these sentences.

accepted	Big Bang	expanding	only	red-shift	shrinking	Steady State

The Big Bang theory is supported by ...., which suggests that the universe is .... However, the .... theory is also supported by this observation. Other evidence has been found which .... supports the .... theory, so this is the currently .... model. Scientists continue to investigate in order to develop and improve established models.

4. a. Complete the sentence.

   There is still a lot about the universe that is not understood, for example, ... energy and dark ....

   b. Cross out the incorrect word from the pair in **bold**.

   Since 1998, observations of supernovae suggest that distant galaxies are receding ever faster. This means that the rate of expansion of the universe is **accelerating / decelerating**. Scientists think that dark energy may explain this. It is not well understood as it cannot be seen and only its effects can be observed.

   c. Complete the sentence.

   Dark mass may explain the rate of rotation of galaxies. Scientists gain a lot of information by observing .... coming from space. Dark mass is not well understood as it cannot be directly .... because it is ....

Sample response

Use these student responses to improve the way you answer this type of question about astronomy. Consider whether the question is answered in enough detail.

Exam-style question

1 Describe the red-shift of light from distant galaxies. **(2 marks)**

Student A *It's when the wavelength of light changes.*

Include the words light, galaxy, spectrum and wavelength in your answer. You could also include what happens to the frequency of the light.

1 a Has student A said how the wavelength changes? **yes / no**

b Has student A described the effect of relative movement? **yes / no**

c Write an improved answer.

....

....

....

....

Exam-style question

2 Explain what the red-shift of light from distant galaxies tells us about them. **(3 marks)**

Student B *The galaxies move fast and expand.*

2 a Highlight the command word in **2**.

b Has student B given a reason or a description? **reason / description**

c Has student B linked their answer to red-shift? **yes / no**

d Complete this table.

What the red-shift of light from distant galaxies tells us	True	False
The universe is expanding		
All galaxies move at the same speed		
The closer a galaxy is to the observer, the faster it moves		
The faster the galaxy moves, the bigger the increase in wavelength		
Red-shift is larger for galaxies that are further away		
The further away a galaxy is from the observer, the faster it moves		

e Use the information from the table to write an improved answer.

Answer the correct command word.

....

....

....

....

Your turn!

It is now time to use what you have learned to answer the exam-style question from page 17. Remember to read the question thoroughly, looking for information that may help you. Make good use of your knowledge from other areas of physics.

Read the exam-style question and answer it using the guided steps below.

Exam-style question

1 **Figure 1** shows the visible spectra for light from the Sun and three distant galaxies, A, B and C.

All the spectra are aligned and to the same scale.

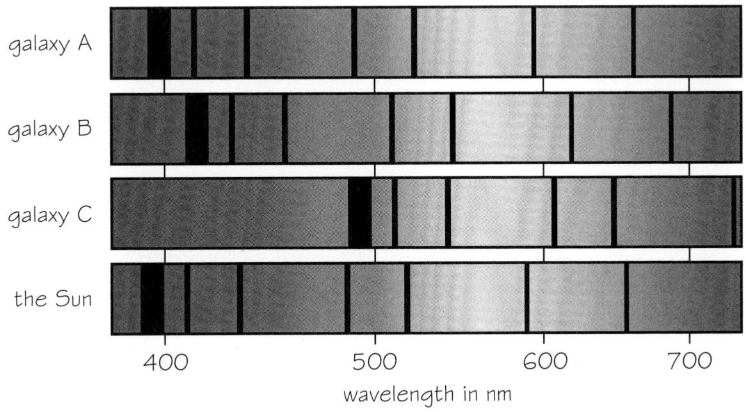

galaxy A

galaxy B

galaxy C

the Sun

400 500 600 700

wavelength in nm

Figure 1

1.1 Explain which galaxy (A, B or C) is nearest to us.

Use **Figure 1** to aid your explanation. **(3 marks)**

1.2 Red-shift supports more than one theory about the origin of the universe.

Explain why the Big Bang theory is the currently accepted model for the origin of the universe. **(2 marks)**

1.3 Scientists are currently discovering things about the universe that they cannot explain yet.

Give **one** example. **(1 mark)**

1.1 i Will the **nearest** galaxy show the most or least red-shift? ....

ii Complete these sentences.

Galaxy .... has the smallest red-shift so galaxy .... has the slowest speed. The galaxy with the slowest speed will be the nearest, so galaxy .... is nearest to us.

An 'explain' answer needs reasons.

Now write your answers to **1.2** and **1.3** below.

Consider the evidence that supports it.

1.2 ....

....

....

....

....

....

....

1.3 ....

Need more practice?

Questions about red-shift are most likely to occur as stand-alone questions, including extended response questions.

Have a go at this exam-style question. You may need to continue your answer on paper.

Exam-style question

1 The table shows two theories used to explain the origin of the universe.

Big Bang theory	Steady State theory
First suggested in the 1920s.	First suggested in 1948.
Around 13.5 billion years ago the universe began from a very small region that was extremely hot and dense. A massive explosion from this single point released huge amounts of radiation. The universe continues to expand from this point.	The universe has always existed.
Gravity caused the matter to come together and form everything in the universe. No new matter is created.	New matter is continuously created within the universe as it expands.

Compare the evidence for the Big Bang and the Steady State theories about the universe.

(4 marks)

In this 'compare' question you need to describe the similarities and the differences between both of the theories. It is not necessary to draw a conclusion.

Boost your grade

Calculations involving speed and distance could be included in exam questions on moving galaxies. Also, you could be asked questions about blue-shift, which is when light comes from something that is moving towards us.

How confident do you feel about each of these **skills**? Colour in the bars.

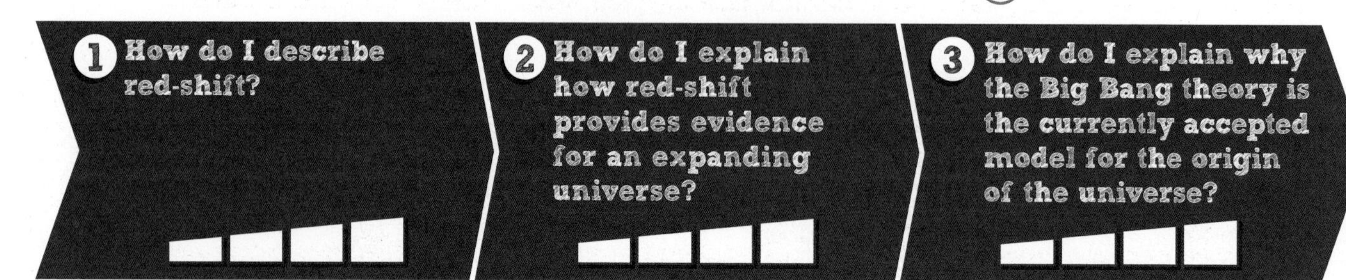

4 Newton's laws, forces and momentum

This unit will help you to understand more about how forces cause acceleration and changes to the movement of objects and how these forces also cause changes to the momentum of objects during impacts.

In the exam you will be asked to answer questions such as the one below.

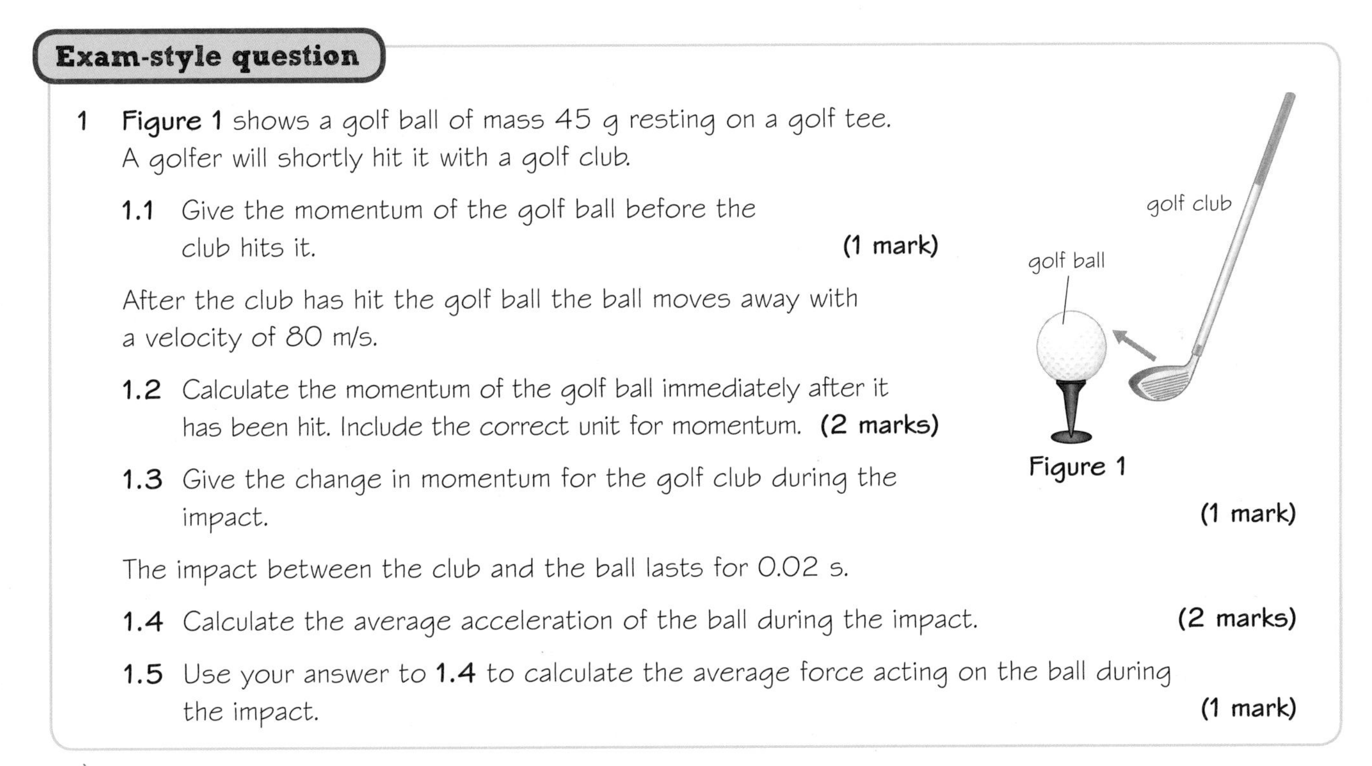

Exam-style question

1 **Figure 1** shows a golf ball of mass 45 g resting on a golf tee. A golfer will shortly hit it with a golf club.

1.1 Give the momentum of the golf ball before the club hits it. **(1 mark)**

After the club has hit the golf ball the ball moves away with a velocity of 80 m/s.

1.2 Calculate the momentum of the golf ball immediately after it has been hit. Include the correct unit for momentum. **(2 marks)**

1.3 Give the change in momentum for the golf club during the impact. **(1 mark)**

The impact between the club and the ball lasts for 0.02 s.

1.4 Calculate the average acceleration of the ball during the impact. **(2 marks)**

1.5 Use your answer to **1.4** to calculate the average force acting on the ball during the impact. **(1 mark)**

Figure 1

You will already have done some work on forces and changes in velocity. Before starting the **skills boosts**, rate your confidence in each area. Colour in the bars.

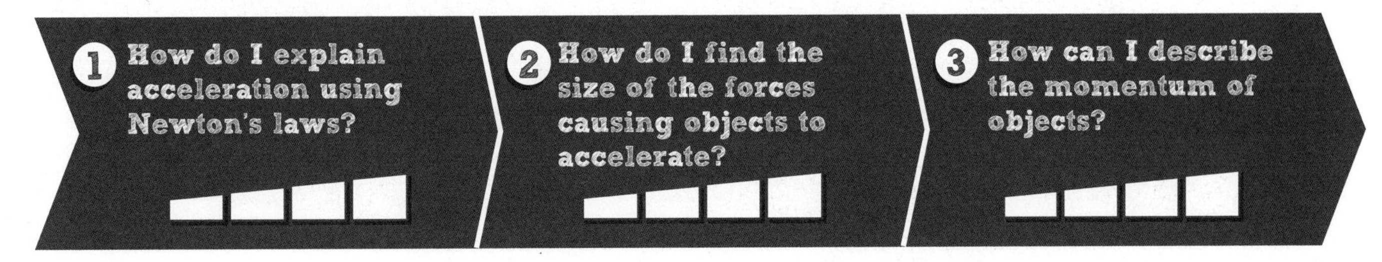

Get started

Resultant forces (sometimes called unbalanced forces) cause change to the movement of objects. This change in movement is called acceleration. An object which is not accelerating can be at rest (stationary), or it can move at a constant speed in a straight line (at a constant velocity).

1. There are many similar words used to describe motion. Draw lines to connect each key word to its correct definition. One has been done for you.

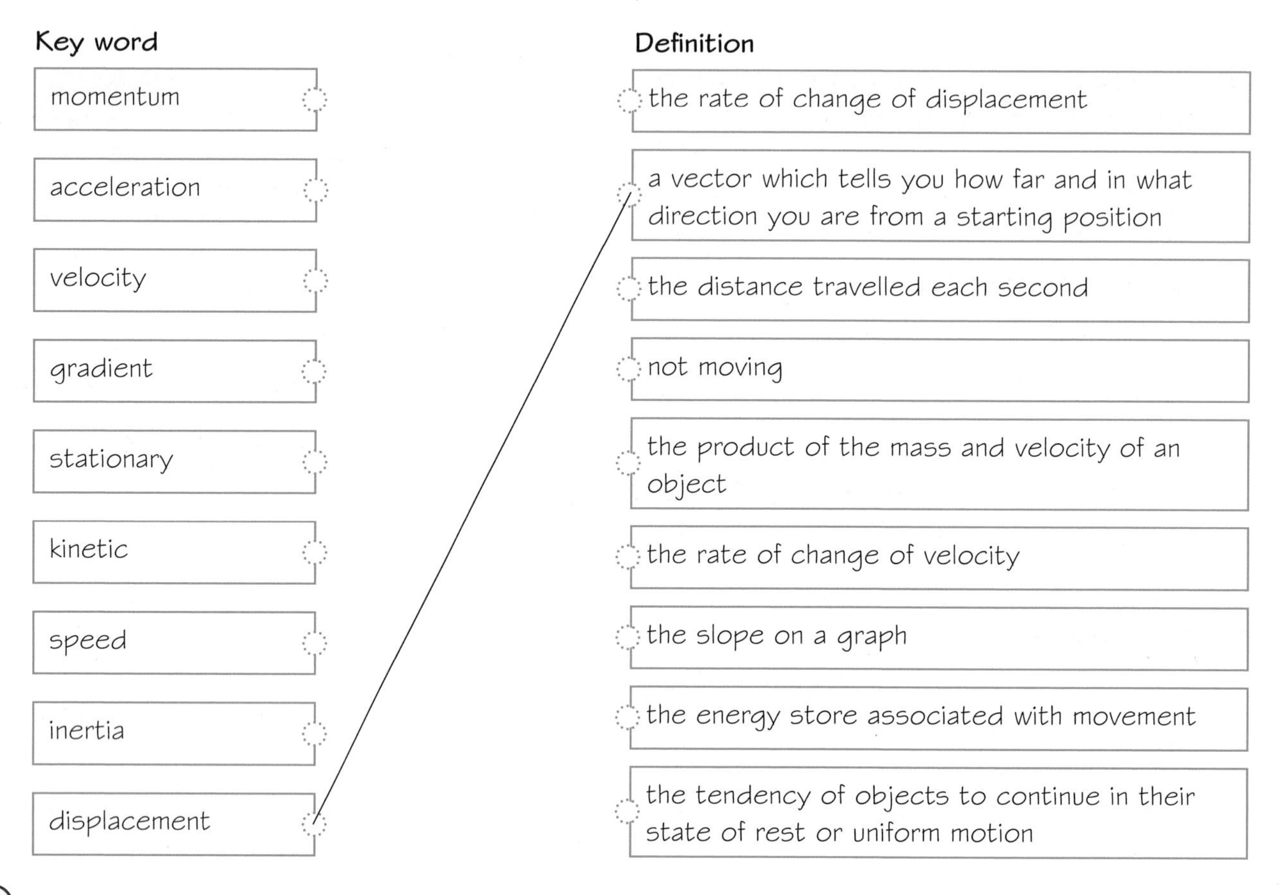

2. The two diagrams below show a book at rest on a table. Each word from the box can be used once, more than once or not at all.

drag	friction	weight	compression	tension	buoyancy	support force

a. On **diagram A** draw and label arrows to show the forces acting **only on the book**. Select words from the box above for the labels.

Remember You are only drawing the forces acting on the book. The book isn't moving so there must be balanced forces acting on it.

b. On **diagram B** draw and label arrows showing the forces acting **only on the table**. Select words from the box above for the labels.

A

B

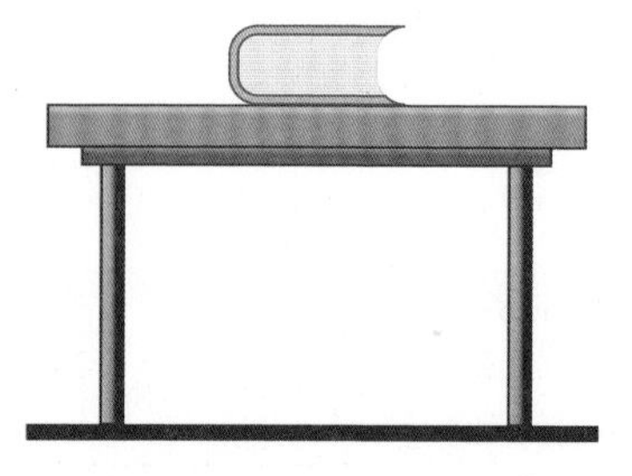

Be careful with your force arrows. They should start where the force acts and point in the direction of the force. The length (*not* thickness) of the arrows should represent the size of the force.

Skills boost

1 How do I explain acceleration using Newton's laws?

Newton gave three laws of motion which describe how forces cause objects to move (or stay stationary). Newton realised that unbalanced forces acting on objects cause all acceleration (changes in velocity). These laws link the **resultant** force on the object to the acceleration that takes place.

1. Complete the sentences about Newton's three laws of motion. Use words from the box below.

You should learn the definitions of Newton's three laws of motion as they are commonly asked for in exams.

proportional	equal	speed	inversely
stationary	mass	opposite	straight

a. If the resultant force acting on an object is zero, a stationary object will remain ....,
whereas a moving object will travel in a .... line at constant ....

b. The acceleration of an object is .... to the resultant force acting on the object and .... proportional to the .... of the object.

c. Whenever two objects interact, the forces they exert on each other are ....
and ....

2. Newton's Second Law allows you to perform calculations linking mass, acceleration and forces. The relationship is usually written as $F = m\,a$.
Draw lines to connect each quantity with its symbol and unit.

Quantity	Symbol	Unit
force	a	newton, N
mass	F	metre per second squared, m/s^2
acceleration	m	kilogram, kg

3. The diagrams in the table below show the forces acting on an object.

a. Draw an arrow to represent the resultant force acting on each of the objects. Write the size of the force. One arrow has been drawn for you.

b. Calculate the acceleration that the object would be experiencing. Write your calculation in the table.

It is important to remember that it is the resultant force which causes acceleration.

To find a resultant force, add all the forces acting in one direction and subtract the forces acting in the opposite direction.

To do this, rearrange equation $F = m\,a$. This equation can be rearranged to $a = F/m$ and $m = F/a$.

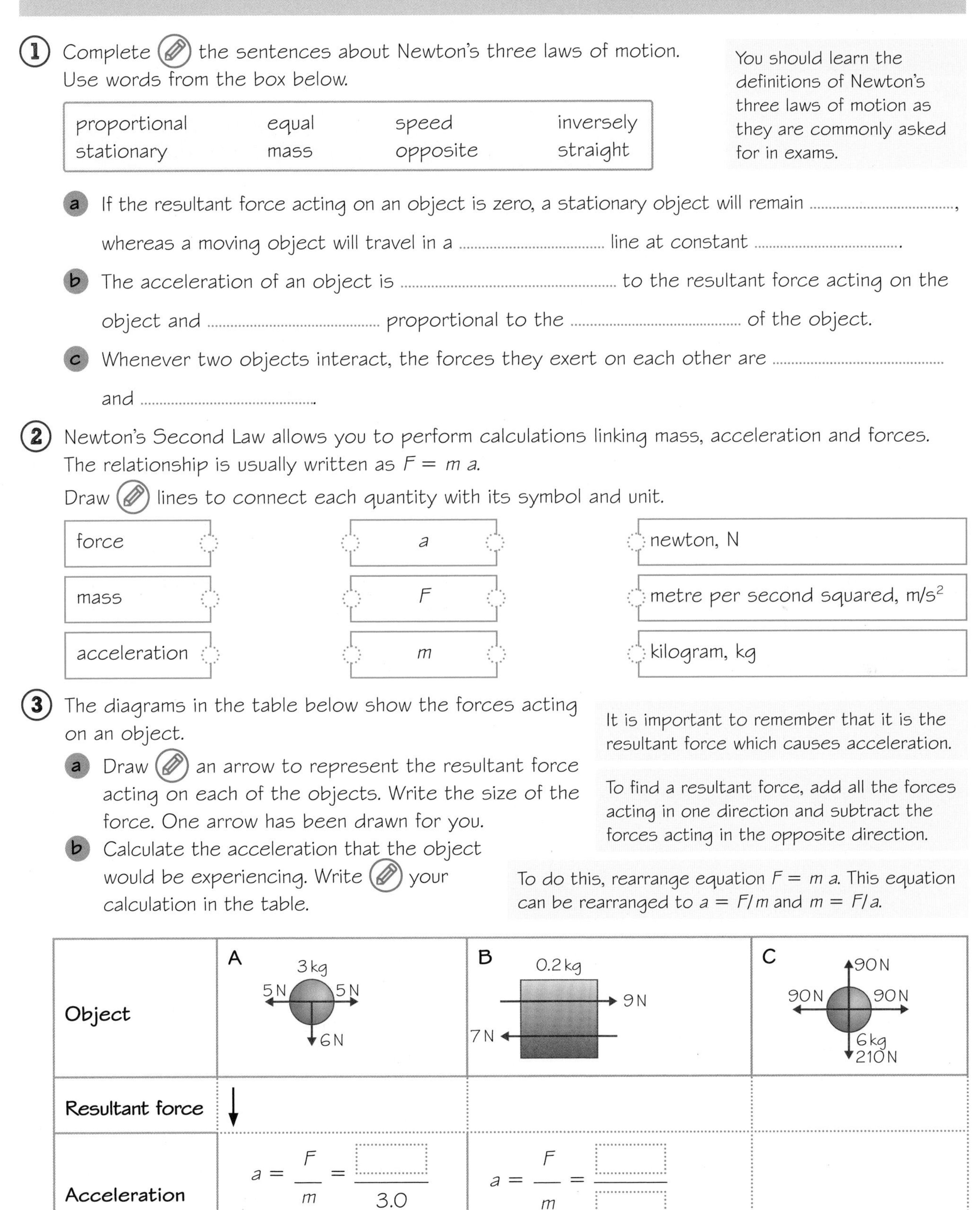

	A	B	C
Object	3 kg; 5 N; 5 N; 6 N	0.2 kg; 9 N; 7 N	90 N; 90 N; 90 N; 6 kg; 210 N
Resultant force	↓		
Acceleration	$a = \frac{F}{m} = \frac{\square}{3.0}$ $a =$ .... m/s^2	$a = \frac{F}{m} = \frac{\square}{\square}$ $a =$	

How do I find the size of the forces causing objects to accelerate?

It is important to be able to find the sizes of the forces causing acceleration, particularly for vehicles. For this you need to be able to use two relationships, one after the other as described here.

Velocity is a vector quantity. This means that it has a size and a direction. When you are calculating changes in velocity (Δv) you need to take into account the direction the object is moving at the start and at the end.

1. Complete this table to calculate the changes in velocity described.

Use the rule that velocity to the right is positive and velocity to the left is negative.

Description of change in velocity	Identifying velocities	Calculation of change in velocity
Change from 4 m/s to the right to 8 m/s to the right	Start velocity = +4 End velocity = +8	Δv = end velocity – start velocity $\Delta v = 8 - 4$ Δv ....
Change from 9 m/s to the right to 3 m/s to the left	Start velocity = +9 End velocity = ...	Δv = end velocity – start velocity $\Delta v = (-3)$ .... $\Delta v =$ ....

To find the **forces** involved when a vehicle changes velocity, use $F = m\,a$. Often the acceleration, a, is not provided, so you need several steps to reach the answer.

2. Complete the steps in the table below.

Stage	Example	Calculation 1	Calculation 2
Underline the start and end velocity; highlight the time.	A motor cycle has a mass of 500 kg. Calculate the force needed to accelerate it from <u>0 m/s to 10 m/s</u> in 5.0 s.	A car has a mass of 1200 kg. Calculate the force needed to accelerate it from 0 m/s to 5.0 m/s in 5.0 s.	A lorry has a mass of 8000 kg. Calculate the force needed to accelerate it from 9.0 m/s to 1.0 m/s in 9.0 s.
Find the change in velocity (Δv). Use the data you have underlined.	0 to 10 m/s $\Delta v = 10 - 0$ $= 10$ m/s	0 to 5.0 m/s	....
Find the acceleration $a = \frac{\Delta v}{t}$	$a = \frac{\Delta v}{t} = \frac{10}{5.0}$ $a = 2.0$ m/s	$a = \frac{\Delta v}{t} = \frac{\square}{\square}$ $a =$	$a = \frac{\Delta v}{t} = \frac{\square}{\square}$ $a =$ ....
Find the size of the forces involved. $F = m\,a$ Use the value of a you have calculated.	$F = m\,a$ $F = 500 \times 2.0$ $F = 1000$ N	$F = m\,a$	.... (to 4 significant figures)

3 How can I describe the momentum of objects?

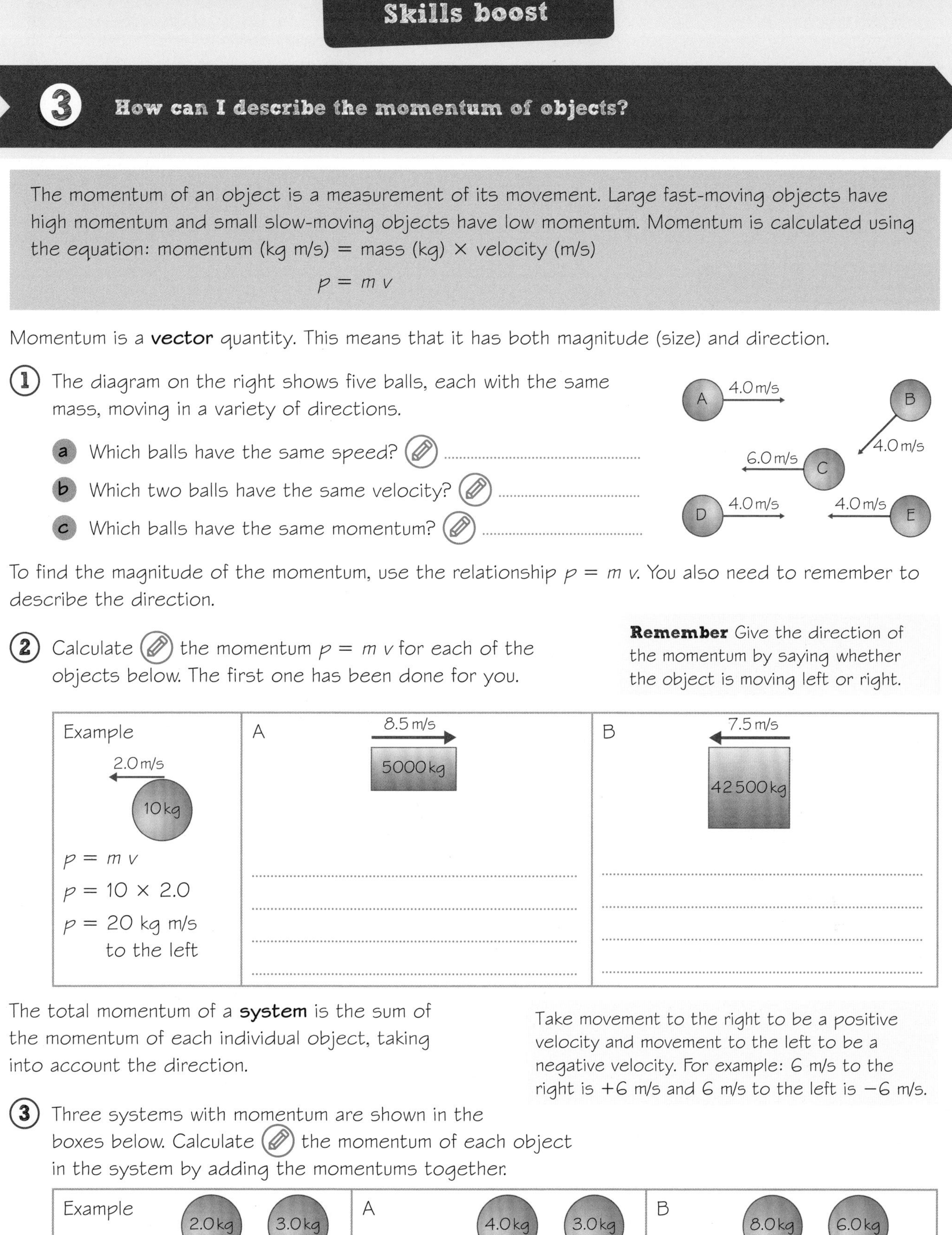

The momentum of an object is a measurement of its movement. Large fast-moving objects have high momentum and small slow-moving objects have low momentum. Momentum is calculated using the equation: momentum (kg m/s) = mass (kg) × velocity (m/s)

$$p = m\,v$$

Momentum is a **vector** quantity. This means that it has both magnitude (size) and direction.

1 The diagram on the right shows five balls, each with the same mass, moving in a variety of directions.

a Which balls have the same speed? ....

b Which two balls have the same velocity? ....

c Which balls have the same momentum? ....

To find the magnitude of the momentum, use the relationship $p = m\,v$. You also need to remember to describe the direction.

2 Calculate the momentum $p = m\,v$ for each of the objects below. The first one has been done for you.

Remember Give the direction of the momentum by saying whether the object is moving left or right.

Example	A	B
2.0 m/s (to the left), 10 kg	8.5 m/s (to the right), 5000 kg	7.5 m/s (to the left), 42 500 kg
$p = m\,v$	....	....
$p = 10 \times 2.0$	....	....
$p = 20$ kg m/s to the left	....	....
	....	....

The total momentum of a **system** is the sum of the momentum of each individual object, taking into account the direction.

Take movement to the right to be a positive velocity and movement to the left to be a negative velocity. For example: 6 m/s to the right is +6 m/s and 6 m/s to the left is −6 m/s.

3 Three systems with momentum are shown in the boxes below. Calculate the momentum of each object in the system by adding the momentums together.

Example	A	B
2.0 kg, 6.0 m/s (to the left); 3.0 kg, 3.0 m/s (to the left)	4.0 kg, 1.5 m/s (to the right); 3.0 kg, 2.0 m/s (to the right)	8.0 kg, 3.0 m/s (to the right); 6.0 kg, 4.0 m/s (to the left)
$p = m\,v$	$p = m\,v$	
2.0 × 6.0 = –12.0 to the left	....	....
3.0 × 3.0 = –9.0 to the left	....	....
–12.0 – 9.0 = –21.0 kg m/s to the left	....	....
	....	....

Sample response

Here are some exam-style questions. Use the student responses to these questions to improve your understanding of how forces cause acceleration and changes in momentum.

Exam-style question

1 A snooker ball has a momentum of 0.30 kg m/s and is travelling at 2.50 m/s.

1.1 Calculate the mass of the ball.

1.2 The ball hits the cushion of the table and bounces off in the opposite direction with a velocity of 2.0 m/s. Calculate the change in momentum of the ball.

Your answer should combine both of the facts given.

Look at this student's answers to the exam-style question.

1.1 $m = \frac{v}{p} = \frac{2.50}{0.30} = 8.3333333$ kg

1.2 change in momentum = change in velocity × mass = 0.5 × 8.333 = 4.17 kg m/s

1 **a** The student has used an incorrect version of the equation linking momentum, mass and velocity for **1.1**. Write the correct version of this equation.

Remember The definition of momentum is from momentum = mass × velocity

b Calculate the correct value.

c What other mistake has the student made in **1.1**?

d What has the student done correctly in **1.2**?

e What mistake has the student made in **1.2**?

Momentum is a vector quantity so take a close look at the velocities.

f Calculate the correct answer to **1.2**.

Your turn!

It is now time to use what you have learned to answer the exam-style question from page 25. Remember to read the question thoroughly, looking for information that might help you. Make good use of your knowledge from other areas of physics.

Exam-style question

1 **Figure 1** shows a golf ball of mass 45 g resting on a golf tee. A golfer will shortly hit the ball with a golf club.

1.1 Give the momentum of the golf ball before the club hits it. **(1 mark)**

....

golf club

golf ball

Figure 1

After the club has hit the golf ball, the ball moves away with a velocity of 80 m/s.

1.2 Calculate the momentum of the golf ball immediately after it has been hit. Include the correct unit for momentum. **(2 marks)**

1.3 Give the change in momentum for the golf club during the impact. **(1 mark)**

If you make a mistake calculating the momentum in **1.1** this won't affect your marks for your answer to **1.2** as you have already been penalised.

....

The impact between the club and the ball lasts for 0.02 s.

1.4 Calculate the average acceleration of the ball during the impact. **(2 marks)**

Use the starting and end velocities and the time of impact. Don't forget to give the correct unit for acceleration.

1.5 Use your answer to **1.4** to calculate the average force acting on the ball during the impact. **(1 mark)**

Use Newton's Second Law here.

Need more practice?

Questions about momentum could occur as part of a question on how objects move or how forces change the velocity or momentum of objects. They could also be part of a question about an experiment or investigation, or as stand-alone questions.

Have a go at this exam-style question.

Exam-style question

1 A cyclist in a race is travelling at 11.0 m/s along a straight section of road. The cyclist sees a crash ahead and brakes suddenly to a speed of 5.0 m/s in a time of 1.5 s.

Which equation links acceleration, change in velocity and time?

1.1 Calculate the average acceleration of the cyclist during braking. **(2 marks)**

The total mass of the cyclist and the bicycle is 60.0 kg.

1.2 Calculate the average braking force the two tyres exert on the road during braking. **(1 mark)**

Use Newton's Second Law.

1.3 Give the size of the force the road exerts on the tyres during braking. **(1 mark)**

Use Newton's Third Law.

....

The cyclist is unable to stop in time and crashes at a speed of 5.0 m/s.

1.4 Calculate the momentum of the cyclist and bicycle just before they crash. **(1 mark)**

Boost your grade

To boost your grade, make sure that you know how to break down calculations into steps. Acceleration links movement and forces so practise using all the equations which involve it.

How confident do you feel about each of these **skills**? Colour in the bars.

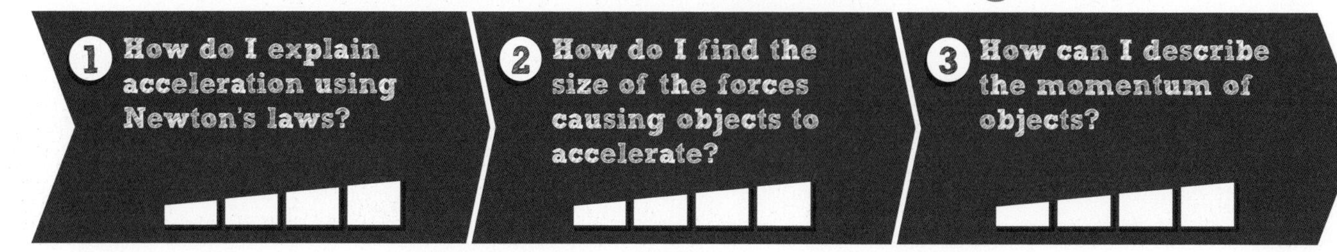

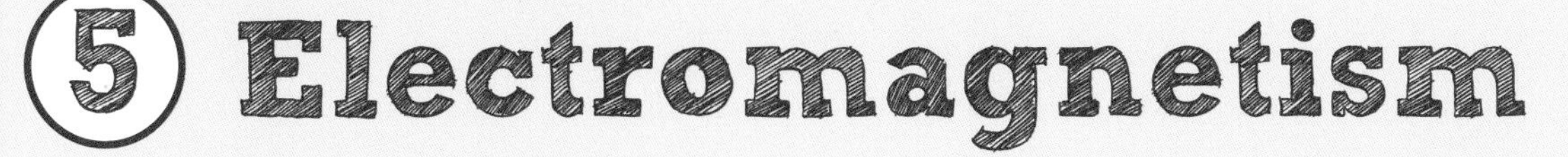

5 Electromagnetism

This unit will help you to understand how electricity and magnetism interact.

In the exam you will be asked to answer questions such as the one below.

Exam-style question

1 A student investigated what happens to a wire in a magnetic field.

Figure 1 shows the equipment the student used.

The wire **XY** is connected to a switch and battery. When the switch is closed, a force acts on the wire.

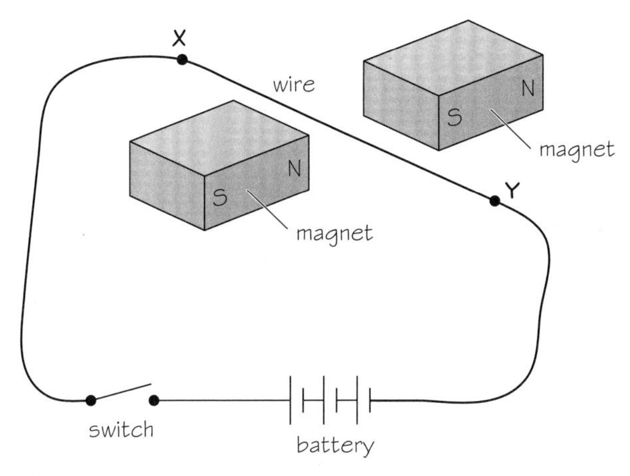

Figure 1

1.1 Explain why a force acts on the wire **XY** when the switch is closed. **(3 marks)**

1.2 Draw an arrow to show the direction of the force acting on the wire **XY**. **(1 mark)**

1.3 Choose the correct equation from the Physics Equations Sheet that links the force acting on a wire carrying an electric current with magnetic flux density and current size. **(1 mark)**

1.4 Describe **three** ways in which the force on the wire can be **increased**. **(3 marks)**

You will already have done some work on electromagnetism. Before starting the **skills boosts**, rate your confidence in each area. Colour in the bars.

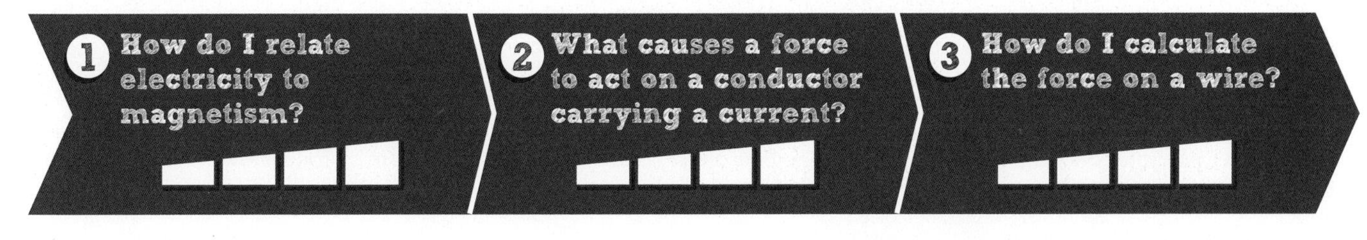

Magnetic field lines are a model used to represent the strength and the direction of magnetic fields. You can investigate magnetic fields using iron filings or plotting compasses. Magnetic field diagrams follow some simple rules:

- Field lines point from north (N) to south (S).
- Field lines cannot touch or cross each other.
- The closer the field lines are to each other, the higher the magnetic flux density (field strength).
- Straight, equally spaced field lines show a uniform field.

1 The diagram on the right represents the magnetic field around a bar magnet.

a Draw an arrow on each field line to show the direction of the magnetic field.

b Write a letter **X** in a region with a strong magnetic field.

c Write a letter **Y** in a region with a weaker magnetic field.

N S

If you place two magnets with unlike poles facing, you get a field diagram between the poles as shown in the diagram below.

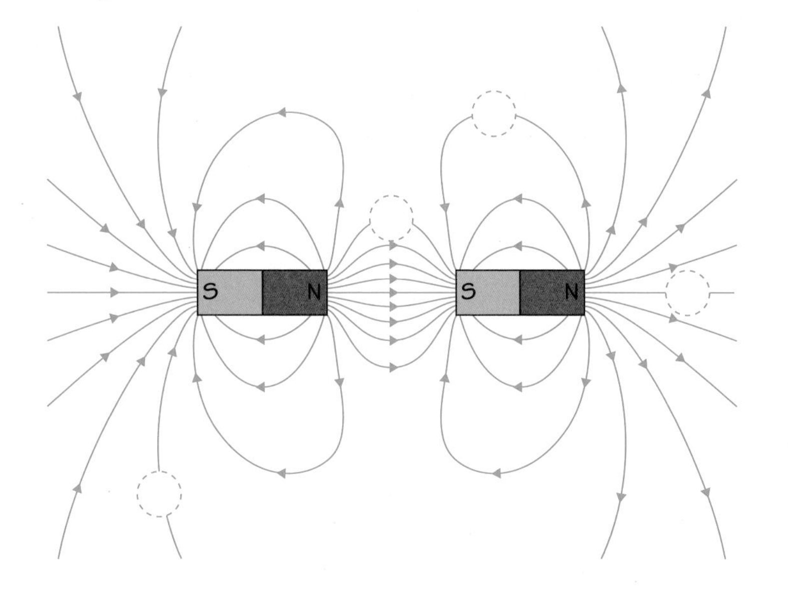

The magnetic fields of the two magnets interact so that there are some field lines between the two magnets.

2 The four circles represent plotting compasses. Add a compass needle to each to show the direction that it would point.

If you place two magnets with wide poles close together, the magnetic field between them looks like the diagram on the right.

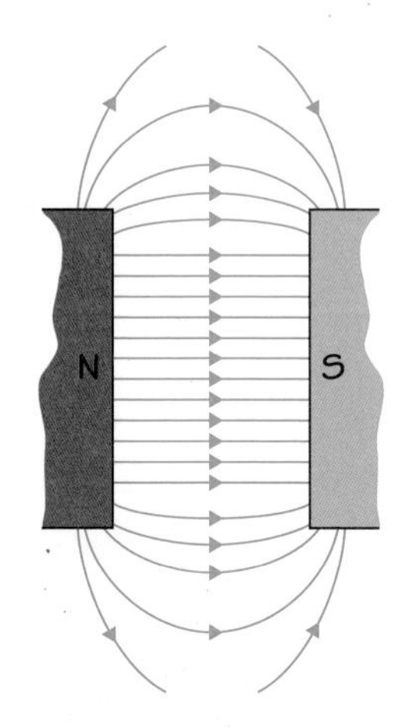

3 How would you describe the magnetic field between the magnets? Tick **two** boxes.

1 How do I relate electricity to magnetism?

A wire carrying an electric current produces a circular magnetic field around itself. You can predict the direction of the magnetic field using the right-hand grip rule.

1 The diagram shows the right-hand grip rule. Complete  this sentence.

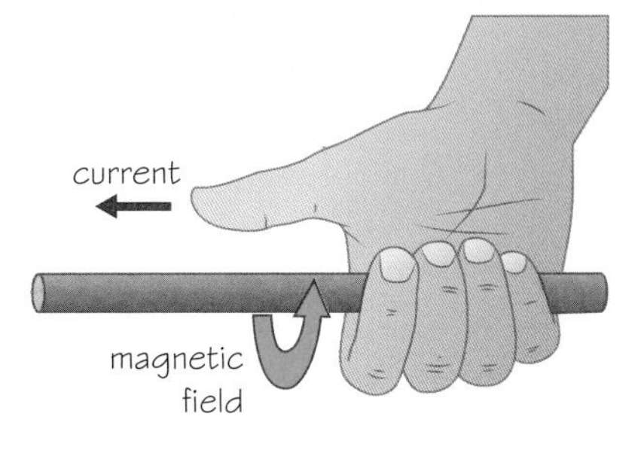

If you were to grip a wire with your right hand, your thumb would point in the direction of the .... and your fingers would point in the direction of the ....

2 The diagram shows the field around a wire carrying a current. The arrows on the field lines show the direction of the magnetic field.
Draw an arrow on the diagram to show the direction of the current.

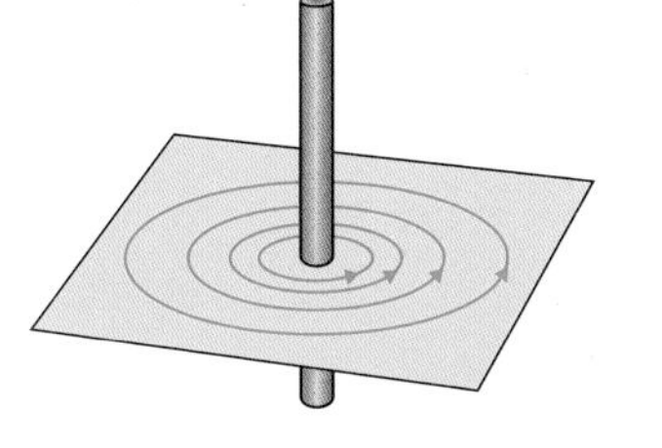

Imagine a right hand gripped around the wire.

The strength of the magnetic field depends on its distance from the wire and the size of the current.

3 Circle the correct words in **bold** to complete the sentences.

Look at the spacing of the field lines.

The magnetic field is **weakest** / **strongest** closer to the wire and gets **stronger** / **weaker** as the distance from the wire increases. The higher the current, the **weaker** / **stronger** the magnetic field.

4 The diagram below shows the magnetic fields around a solenoid and a bar magnet.
Tick the correct boxes to show which statements are true and which are false.

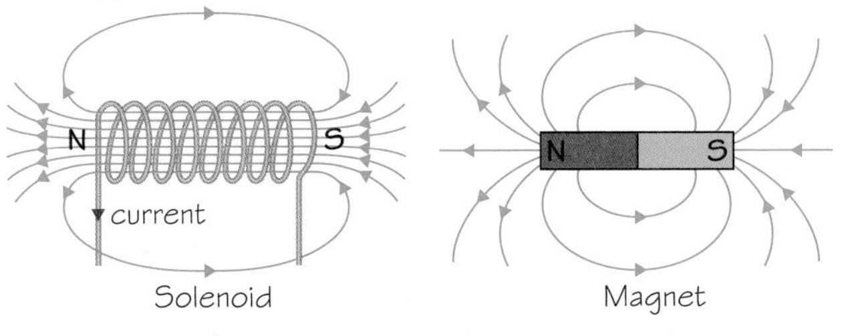

Solenoid   Magnet

Remember The closer the field lines, the greater the magnetic field strength (flux density).

Statement	True	False
The magnetic fields have the same basic shape.		
The magnetic field strength (magnetic flux density) at the poles of the magnet is greater than the field strength at the poles of the solenoid.		
The magnetic field strength (magnetic flux density) at the sides of the magnet is greater than the field strength at the sides of the solenoid.		

2 What causes a force to act on a conductor carrying a current?

The magnetic field around a wire carrying an electric current can interact with other magnetic fields. For example, it can interact with the magnetic field produced by a permanent magnet. These interactions produce forces that may cause the wire or magnet to move.

1 This diagram shows the magnetic field produced when a wire carrying a current is put between two magnets.

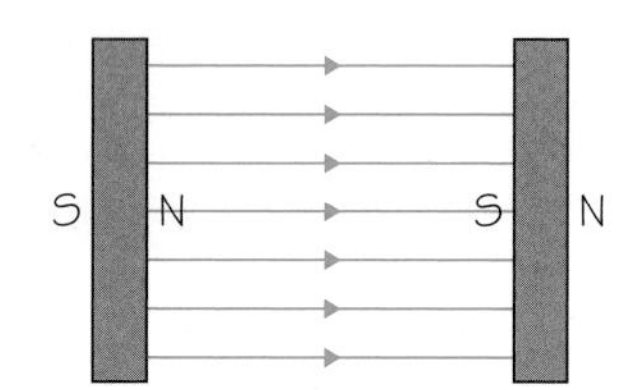

Two flat magnets produce a uniform magnetic field between them.

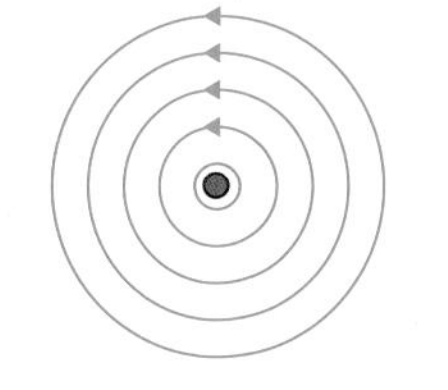

A magnetic field goes around a wire carrying a current.

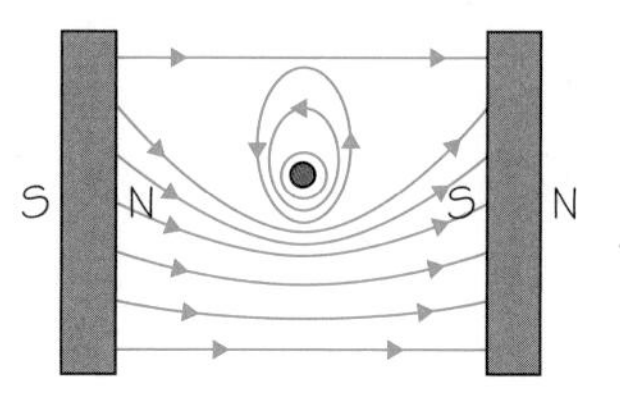

When the wire carrying a current is put between the magnets, the two fields interact to produce a force.

a Write a letter **X** on the right-hand diagram where the magnetic field is strongest.

b Write a letter **Y** on the right-hand diagram where the field is weakest.

c Draw an arrow on the right-hand diagram to show the direction of the force acting on the wire.

To predict the direction of the force on the wire, visualise the field lines as elastic bands. Imagine the elastic bands are trying to straighten by catapulting the wire away.

The wire also exerts a force on the magnets (Newton's Third Law).

d Circle the correct words in **bold** to complete the sentences.

> The force of the wire acting on the magnets is in the **same / opposite** direction to the force of the magnets acting on the wire.
> The size of the force of the wire acting on the magnets is **the same as / different to** the size of the force of the magnets acting on the wire.

To create the greatest force, the magnetic field from the magnets must be at right-angles to the current in the wire. Fleming's left-hand rule (shown in the diagram) gives an easy way to predict the direction of the force (and so the direction of movement) of a wire carrying an electric current.

thuMb: Movement
Forefinger: Field (N to S)
seCond finger: Current (+ to –)

2 The diagram below shows three wires in three different magnetic fields. The arrows show the direction of the current flowing in each wire.

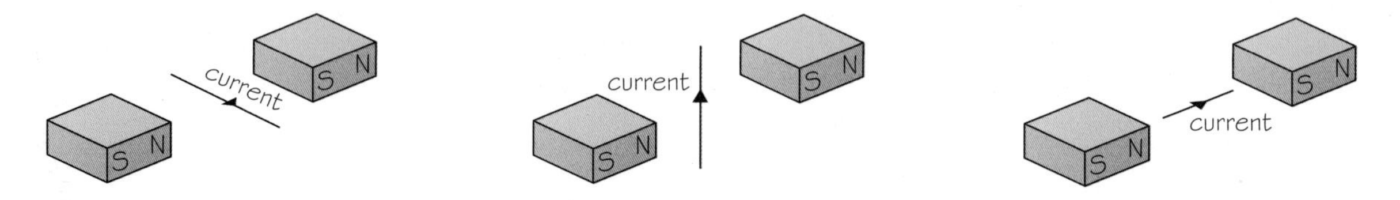

Draw arrows to show the direction of the force acting on each wire. If no force is acting, label the wire with the words 'no force'.

Skills boost

3 How do I calculate the force on a wire?

The force acting on a wire in a magnetic field depends on the strength of the magnetic field, the size of the current and the length of the wire carrying the current. Remember that the force is greatest when the magnetic field and wire are at right-angles to each other.

You can calculate the size of the force acting on a wire using this equation:

force (N) on a conductor (at right-angles to a magnetic field) carrying a current
= magnetic flux density (tesla, T) × current (A) × length (m)

$F = B\,I\,l$

You don't need to learn this equation, but you do need to use it correctly.

1 Draw lines to link each physical quantity to its symbol and unit. One has been done for you.

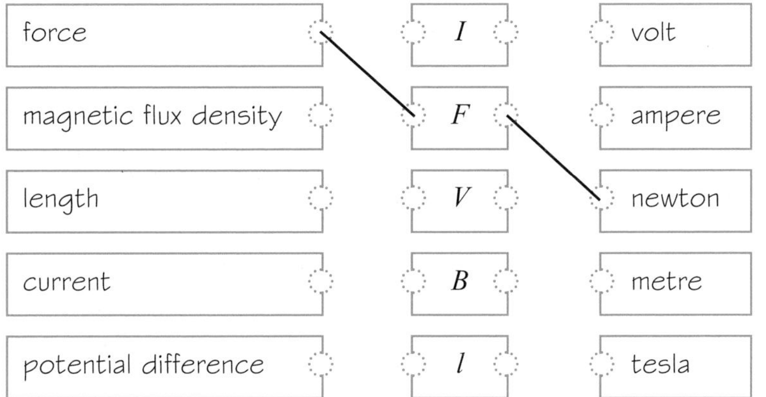

You need to learn the physical quantities, their symbols and their units.

Not all physical quantities have symbols that match the first letter of the physical quantity.

2 The diagram shows two magnets producing a strong uniform magnetic field. The north pole of each magnet is shown in red. The yoke is a frame to hold the magnets.

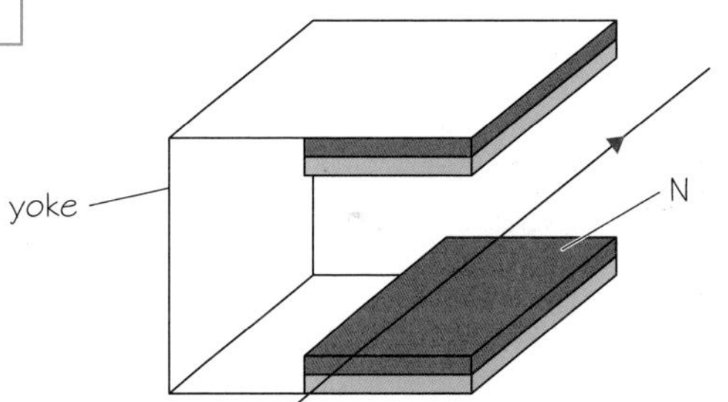

a Draw an arrow to show the direction of the magnetic field and label it 'B'.

b Draw an arrow from the wire to show the direction of the force acting on the wire.

Use Fleming's left-hand rule to predict the direction of the force acting on the wire.

c Draw an arrow from the yoke to show the direction of the force acting on the yoke.

Remember Apply Newton's Third Law here.

d Give **two** ways in which the force acting on the wire could be doubled in size.

$F = B\,I\,l.$

....

....

3 The diagram shows a wire carrying a current in a magnetic field. A 12 cm length of wire in a magnetic field of flux density 0.2 T carries a current of 1.5 A. Calculate the force acting on the wire using the equation: $F = B\,I\,l$.

S N S N I

....

....

....

Check the units.

Sample response

Look at these exam-style questions and answers given by a student

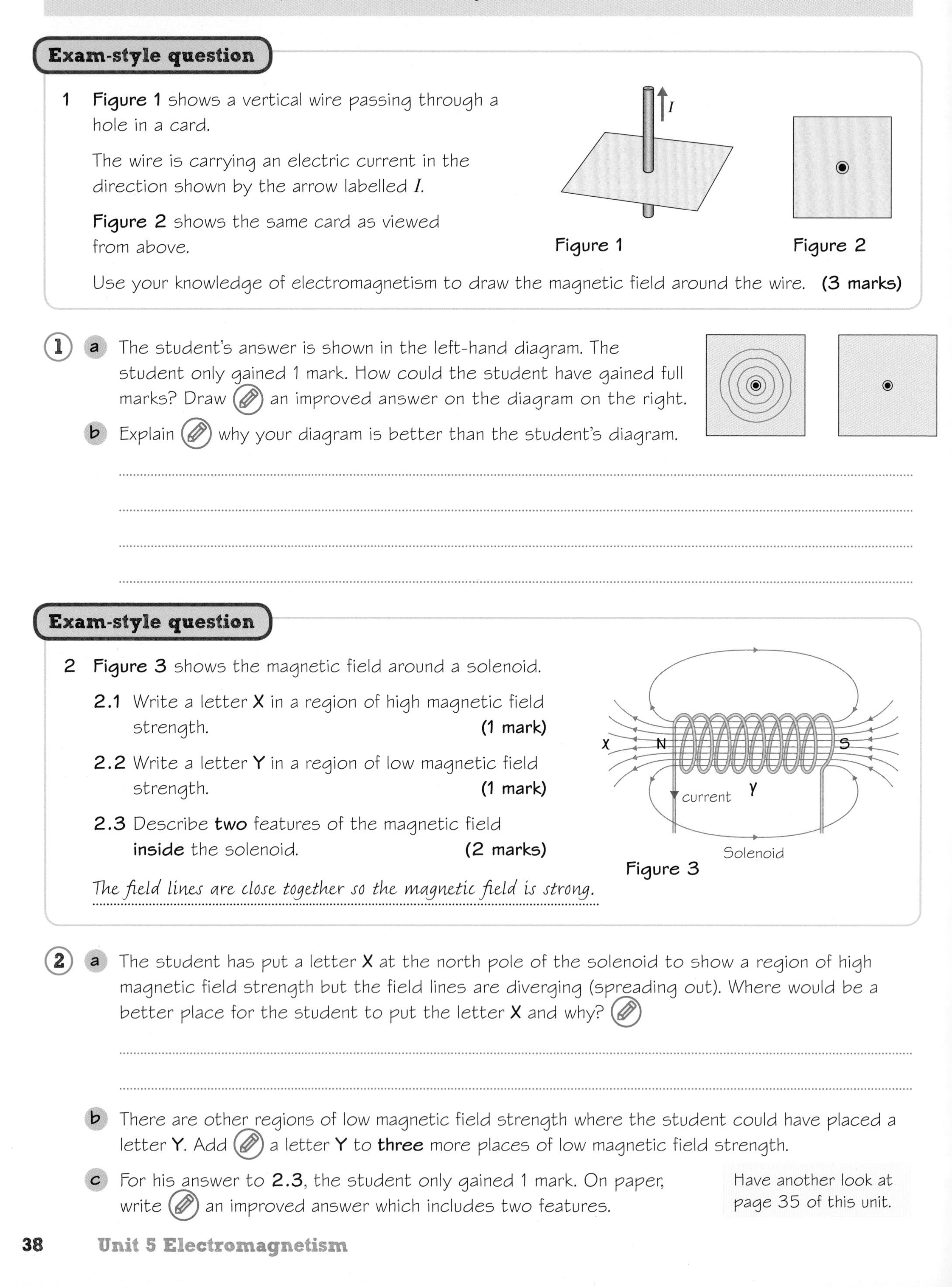

Exam-style question

1 **Figure 1** shows a vertical wire passing through a hole in a card.

The wire is carrying an electric current in the direction shown by the arrow labelled *I*.

Figure 2 shows the same card as viewed from above.

Figure 1

Figure 2

Use your knowledge of electromagnetism to draw the magnetic field around the wire. **(3 marks)**

① **a** The student's answer is shown in the left-hand diagram. The student only gained 1 mark. How could the student have gained full marks? Draw an improved answer on the diagram on the right.

b Explain why your diagram is better than the student's diagram.

....

....

....

....

Exam-style question

2 **Figure 3** shows the magnetic field around a solenoid.

2.1 Write a letter **X** in a region of high magnetic field strength. **(1 mark)**

2.2 Write a letter **Y** in a region of low magnetic field strength. **(1 mark)**

2.3 Describe **two** features of the magnetic field **inside** the solenoid. **(2 marks)**

The field lines are close together so the magnetic field is strong.

Figure 3

② **a** The student has put a letter **X** at the north pole of the solenoid to show a region of high magnetic field strength but the field lines are diverging (spreading out). Where would be a better place for the student to put the letter **X** and why?

....

....

b There are other regions of low magnetic field strength where the student could have placed a letter **Y**. Add a letter **Y** to **three** more places of low magnetic field strength.

c For his answer to **2.3**, the student only gained 1 mark. On paper, write an improved answer which includes two features.

Have another look at page 35 of this unit.

Your turn!

It is now time to use what you have learned to answer the exam-style question from page 33. Remember to read the question thoroughly, looking for information that may help you. Make good use of your knowledge from other areas of physics.

Remember Use Fleming's left-hand rule when answering this question.

Exam-style question

1 A student investigated what happens to a wire in a magnetic field.

Figure 1 shows the equipment the student used.

The wire **XY** is connected to a switch and battery. When the switch is closed, a force acts on the wire.

X
wire
S N
magnet
S N
magnet
Y
switch
battery

Figure 1

1.1 Explain why a force acts on the wire **XY** when the switch is closed. **(3 marks)**

....

....

....

....

....

....

There are three marks available for **1.1**. Think about three important points to write about – use **Figure 1** to help you.

1.2 Draw an arrow to show the direction of the force acting on the wire **XY**. **(1 mark)**

Work out the direction of the force acting on the wire using Fleming's left-hand rule.

1.3 Choose the correct equation from the Physics Equations Sheet that links the force acting on a wire carrying an electric current with magnetic flux density and current size. **(1 mark)**

....

....

....

1.4 Describe **three** ways in which the force on the wire can be **increased**. **(3 marks)**

....

....

....

....

....

Use the equation to help you. Think about which factors on the right side of the equation could increase the force acting on the wire.

Need more practice?

Questions about electromagnetism could occur as part of a question about an experiment or investigation about devices that rely on an electromagnet to work, or as stand-alone questions.

Have a go at this exam-style question.

Exam-style question

1 A student measured the magnetic flux density produced between two magnets.

The student placed the two magnets on a top-pan balance as shown in **Figure 1**.

The top-pan balance reading changed by 16.0 g when the student passed a current of 3.2 A through the wire and placed it in the magnetic field.

Figure 1

1.1 Calculate the force acting on the balance.

Use gravitational field strength (g) = 10 N/kg. **(3 marks)**

Remember Weight (newton, N) = mass (kilogram, kg) × gravitational field strength (newton per kilogram, N/kg), $W = m\ g$.

1.2 Calculate the magnetic flux density produced between the two magnets, assuming that 0.10 m of wire is in the magnetic field. Use an equation from the Physics Equations Sheet. **(4 marks)**

Boost your grade

Make sure you can use select and apply the equation: force on a conductor (at right-angles to a magnetic field) carrying a current (newton, N) = magnetic flux density (tesla, T) × current (ampere, A) × length (metre, m): $F = B\ I\ l$.

How confident do you feel about each of these **skills**? Colour in the bars.

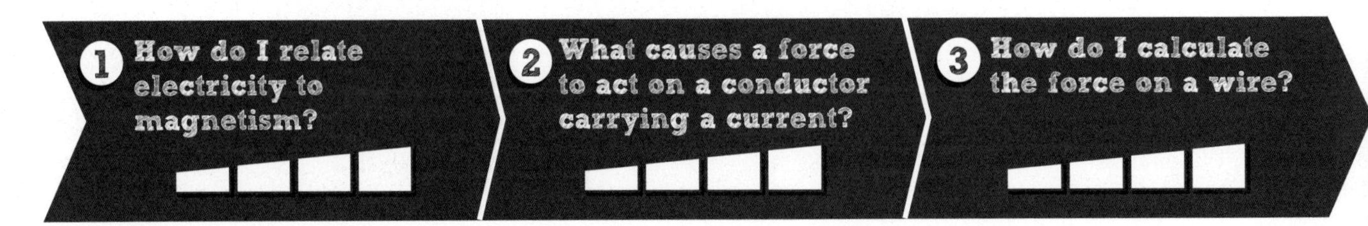

6 Analysing energy transfers

This unit will help you learn more about energy transfers. It will show you how to use an energy analysis to make predictions about what can and cannot happen.

In the exam, you will be asked to answer questions such as the one below.

Exam-style question

1 **Figure 1** shows a bicycle pump inflating a tyre.

1.1 Describe the energy transfer that takes place as the pump compresses the air particles. **(2 marks)**

The bicycle pump piston has an area of $0.125\,m^2$. The pressure of the particles increases by 1.2 Pa with every push.

1.2 Calculate the work done when the pump piston moves down 0.3 m. **(3 marks)**

1.3 The temperature of the pumped air rises from 293 K to 328 K. Explain this rise in temperature in terms of particle movement. **(1 mark)**

1.4 The volume of the air at the start is $25\,cm^3$ with a pressure of 1.4×10^5 Pa.
After the pump is pressed down, the volume is $7\,cm^3$.

Calculate the new pressure when the air in the pump returns to its original temperature.

Use the correct equation from the Physics Equations Sheet. **(2 marks)**

Figure 1

Figure 2

You will already have done some work on energy and energy transfer. Before starting the **skills boosts**, rate your confidence for each skill. Colour in the bars.

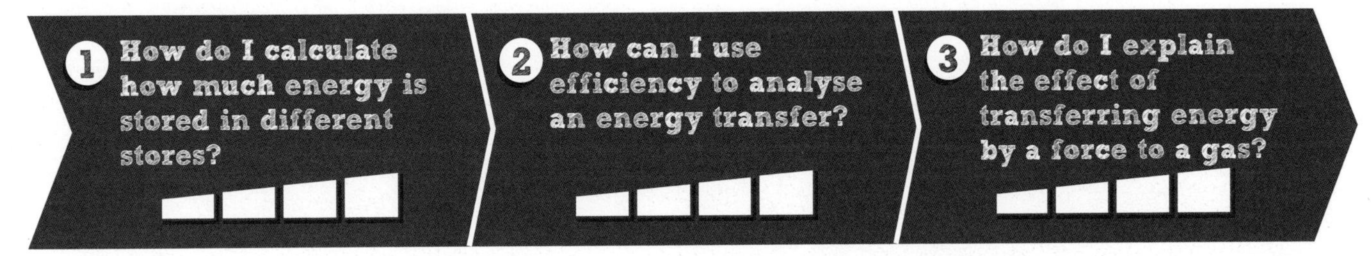

Get started

Energy is transferred by three different processes: forces doing work, electric currents doing work, and heating effects. These processes cause one store of energy to decrease while other stores increase.

The energy decrease in one store always matches the energy increase in other stores, so the total amount of energy is always conserved.

1 Write the process that causes energy to be transferred when:

Try not to use the term 'electricity' as it is too general. It is electric current that transfers energy.

a a box is pushed along the floor ....

b an ice cube is placed into a cup of hot tea ....

c a lamp lights up in a circuit ....

....

2 Circle the correct words in bold to describe the energy transfers.

In a battery-operated circuit, energy is transferred to an electric motor by **an electric current / a force / heating**. The motor lifts a load through a distance and transfers energy to the load by **an electric current / a force / heating**. It also causes an increase in temperature by **an electric current / a force / heating**.

3 Complete the table to describe the changes in energy stores when energy transfer processes happen.

When you describe a store, don't just say 'chemical store'. Make sure you say what or where that store is. For example, 'chemical store of the battery'.

Process	Store that decreases	Stores that increase
using a battery-powered electric motor to lift an object	chemical store of the battery	gravitational potential store of the object being lifted thermal stores of the surroundings
an object falling from a great height		kinetic store of the object falling thermal stores of the surroundings
a rubber dinghy being inflated using a hand pump	chemical store in the person doing the work	
a kettle being used to heat water	chemical store at a power station (coal)	
an arrow being launched from a bow		

4 Use the principle of conservation of energy to complete these sentences.

Remember The total energy after the transfer is equal to the total energy before the transfer.

a A student uses an electric heater to warm up a sample of water. The heater is provided with 4000 J of energy by the electric current. The thermal store of the water increases by 3500 J and .... J is wasted by heating the surroundings.

b A lorry burns fuel with an energy content of 4.0 kJ and its kinetic store increases by 1.5 kJ. The thermal store of the truck and surroundings increases by .... kJ.

c A battery provides .... J of energy through an electric current that powers a motor. The motor lifts with a force that transfers 500 J to a gravitational store but also causes 50 J of heating.

How do I calculate how much energy is stored in different stores?

To calculate the changes in amounts of energy in a store, or the energy transferred through work done, you need to measure the physical properties that change. For example, the energy stored in a spring increases when you stretch it, so you measure the change in length.

You also need to understand the factors that affect the size of each energy store or energy transfer so that you can select the right equation to use.

1 For each energy store or energy transfer, draw a line to the equation you need to use to calculate the amount of energy stored or transferred.

Work done is the way in which forces transfer energy to objects. For example, when you push a bicycle pump handle, you transfer energy from your body to the particles of air in the pump.

You need to recall the equations marked with a * because they won't be given to you on the Physics Equation Sheet.

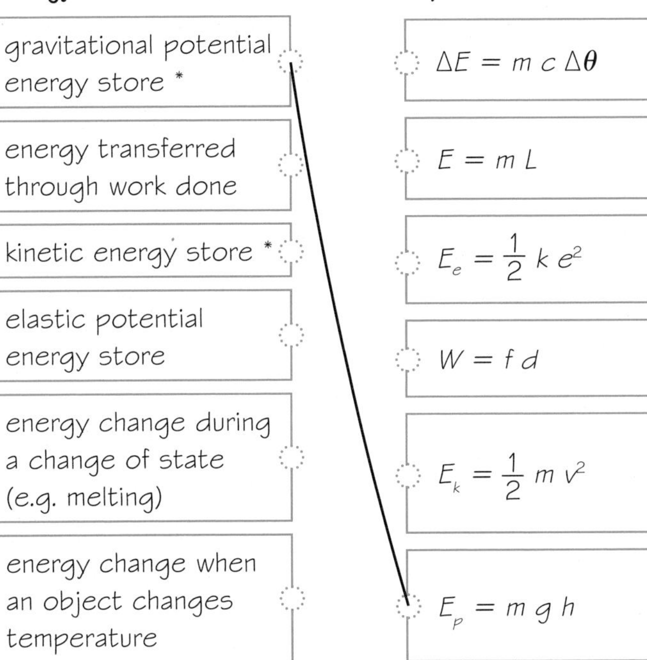

Remember

Symbol		Unit
m	mass	kg
h	height	m
c	specific heat capacity	J/kg °C
$\Delta\theta$	change in temperature	°C
k	spring constant	N/m
e	extension	m
L	specific latent heat	J/kg
v	speed	m/s
g	gravitational field strength	N/kg
W	work done	J
f	force	N

2 Follow the method shown in the example to calculate the energy.

	Example	Calculation
a Highlight the key data.	Calculate the energy change when a 4.00 kg block of aluminium with specific heat capacity of 902 J/kg °C is heated so that its temperature rises from 5.0 °C to 40.0 °C.	Calculate the energy change when a remote-controlled toy car of mass 3.0 kg accelerates from rest (0 m/s) to 2.0 m/s.
b Select the correct energy equation.	$\Delta E = m\,c\,\Delta\theta$	....
c Substitute in the values you have highlighted.	$\Delta E = 4.00 \times 902 \times (40 - 5)$	....
d Calculate the answer.	$\Delta E = 126\ 280$	....
e Choose the correct number of significant figures and add the units.	$\Delta E = 126$ kJ	....

2 How can I use efficiency to analyse an energy transfer?

The efficiency of an energy transfer tells you how much energy a useful process transfers and how much energy is transferred in a way that is not useful.

There are two versions of the efficiency equation, both of which you need to memorise. The one you need to use here is:

$$\text{efficiency} = \frac{\text{useful output energy transfer}}{\text{total input energy transfer}}$$

The energy transferred by the process we want to happen is **useful energy**.

The energy transferred by a process we don't want to happen is **wasted energy**.

1. Read this question, and answer it using the guided steps below.

A toy uses a spring to launch sponge darts through the air.

The spring used to launch the darts has a spring constant of 200 N/m.

When a dart is launched, the spring is compressed by 0.50 m. The efficiency of the spring launcher is 0.90.

Calculate the **kinetic energy** of the dart.

a. Show the important data in the question.

   i Circle the two pieces of data that allow you to calculate the energy stored in the spring.

   ii Underline the efficiency of the energy transfer.

b. Circle the equation that allows you to calculate the energy stored in the spring.

$\Delta E = m\,c\,\Delta\theta$ $\quad E_e = \frac{1}{2}\,k\,e^2$ $\quad W = f\,d$ $\quad E_k = \frac{1}{2}\,m\,v^2$ $\quad E_p = m\,g\,h$

Remember k represents the spring constant and e the extension or compression.

c. Use the equation to calculate the energy stored in the spring when it is compressed.

d. Use the efficiency equation to calculate the energy transferred to the dart by the spring during a launch.

3 How do I explain the effect of transferring energy by a force to a gas?

Inside a sealed balloon, there are many air particles. Each air particle hits the inside walls of the balloon with a force. The huge number of collisions gives the balloon its pressure.

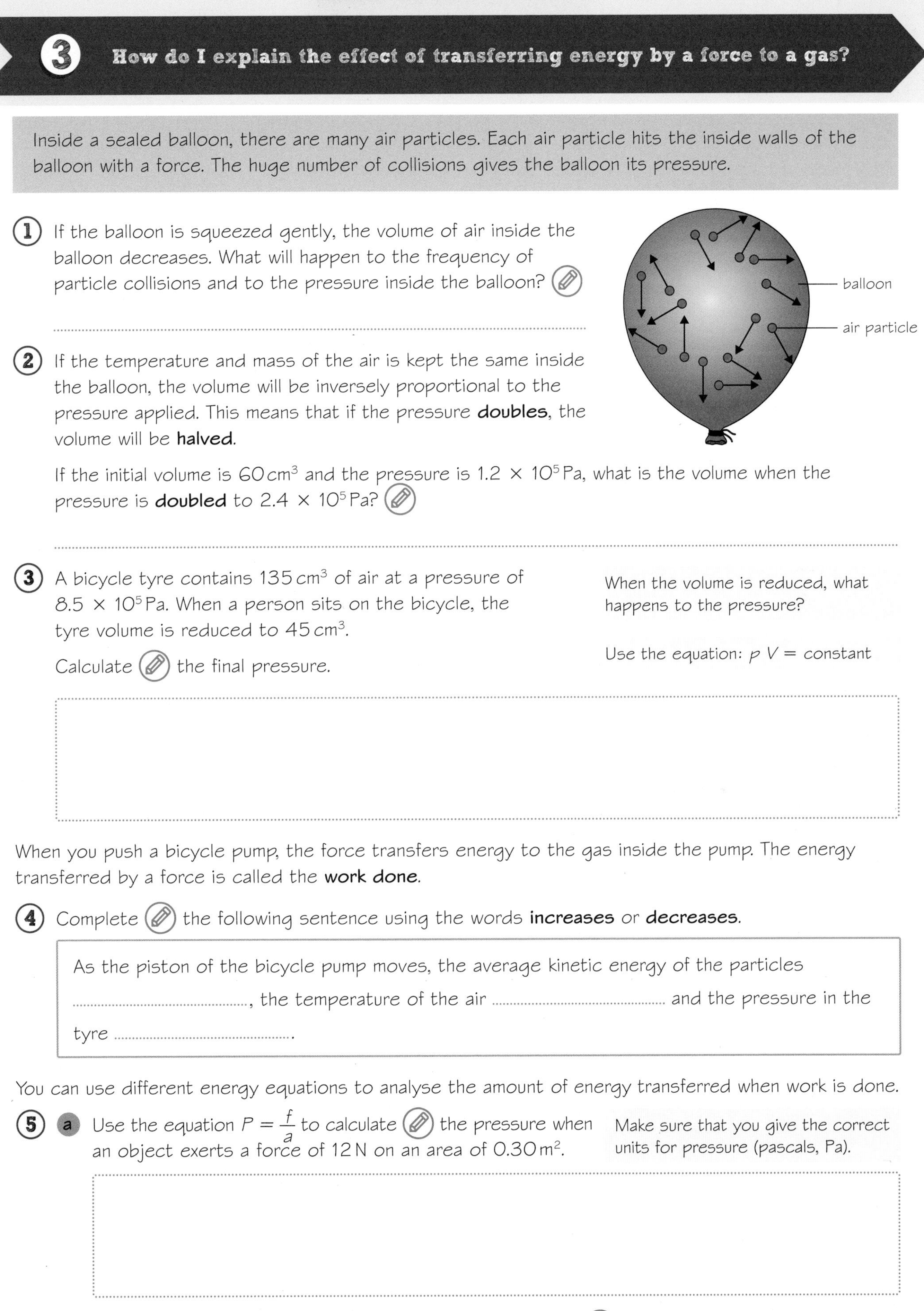

1 If the balloon is squeezed gently, the volume of air inside the balloon decreases. What will happen to the frequency of particle collisions and to the pressure inside the balloon?

....

2 If the temperature and mass of the air is kept the same inside the balloon, the volume will be inversely proportional to the pressure applied. This means that if the pressure **doubles**, the volume will be **halved**.

If the initial volume is $60\,cm^3$ and the pressure is $1.2 \times 10^5\,Pa$, what is the volume when the pressure is **doubled** to $2.4 \times 10^5\,Pa$?

....

3 A bicycle tyre contains $135\,cm^3$ of air at a pressure of $8.5 \times 10^5\,Pa$. When a person sits on the bicycle, the tyre volume is reduced to $45\,cm^3$.

Calculate the final pressure.

When the volume is reduced, what happens to the pressure?

Use the equation: $p\,V$ = constant

When you push a bicycle pump, the force transfers energy to the gas inside the pump. The energy transferred by a force is called the **work done**.

4 Complete the following sentence using the words **increases** or **decreases**.

As the piston of the bicycle pump moves, the average kinetic energy of the particles ...., the temperature of the air .... and the pressure in the tyre ....

You can use different energy equations to analyse the amount of energy transferred when work is done.

5 **a** Use the equation $P = \frac{f}{a}$ to calculate the pressure when an object exerts a force of 12 N on an area of $0.30\,m^2$.

Make sure that you give the correct units for pressure (pascals, Pa).

b Use the equation for work done to $W = f\,d$ to calculate the energy transferred when a force of 12 N is moved over a distance of 1.2 m.

....

Sample response

Here are some exam-style questions. Use the student responses to these questions to improve your understanding of describing and calculating energy changes.

Exam-style question

1 A coconut of mass 1.0 kg falls from a height of 3.6 m.

Calculate the maximum speed that the coconut will reach before hitting the ground.

The gravitational field strength is 10 N/kg.

Gravitational potential energy = m h = 1.0 × 3.6 = 3.6

Velocity $v = \frac{2E}{m} = \frac{2 \times 3.6}{1.0} = 7.2\ m/s$

(3 marks)

1 a The student has used the wrong equation for gravitational potential energy.

Write the correct equation and calculate the correct value.

b The student has also rearranged the equation linking velocity to kinetic energy incorrectly.

Use the correct arrangement of the equation to calculate the velocity.

Exam-style question

2 An electric kettle is provided with 30 000 J of energy by an electric current.

The heating element has an efficiency of 0.90 and is used to heat 0.5 kg of water.

Water has a specific heat capacity of 4200 J/kg °C.

2.1 Calculate the change in temperature for the water during this heating process.

Temperature change = 14.3 °C

(3 marks)

The student has not shown their working. They have missed out a stage and so their answer is incorrect. If you show correct working you will gain some marks, even if your final answer is incorrect.

2 a What is the value of the energy transferred to the water taking into account the efficiency of the transfer? ....

b What is the correct relationship between temperature rise, specific heat capacity, mass and energy transferred?

$\Delta\theta =$

c What is the correct unit for temperature change? ....

d What is the correct answer to **2.1**?

Your turn!

It is now time to use what you have learned to answer the exam-style question from page 41.

Remember to read the question thoroughly, looking for information that may help you.

Make good use of your knowledge from other areas of physics.

Exam-style question

1 **Figure 1** shows a bicycle pump inflating a tyre.

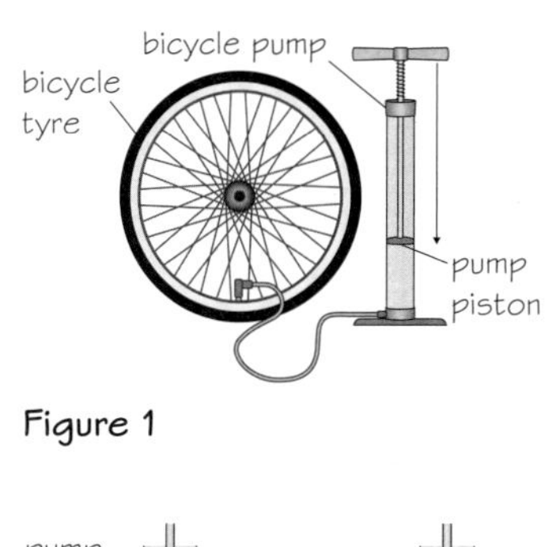

Figure 1

1.1 Describe the energy transfer that takes place as the pump compresses the air particles. **(2 marks)**

....

....

The bicycle pump piston has an area of 0.125 m^2. The pressure of the particles increases by 1.2 Pa with every push.

1.2 Calculate the work done when the pump piston moves down 0.3 m. **(3 marks)**

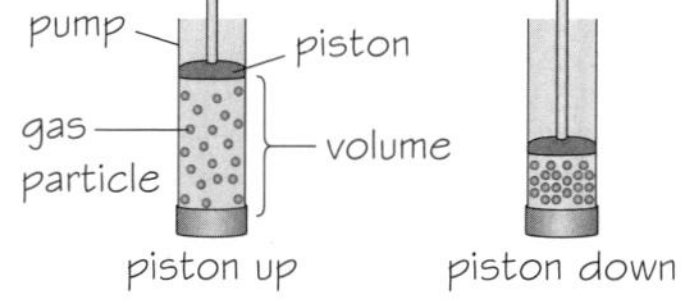

Figure 2

Make sure that you include the units for work done.

1.3 The temperature of the pumped air rises from 293 K to 328 K. Explain this rise in temperature in terms of particle movement. **(1 mark)**

....

....

1.4 The volume of the air at the start is 25 cm^3 with a pressure of 1.4×10^5 Pa. After the pump is pressed down, the volume is 7 cm^3.

Use the equation: pV = constant

Calculate the new pressure when the air in the pump returns to its original temperature.

Use the correct equation from the Physics Equations Sheet. **(2 marks)**

Need more practice?

Questions about energy transfer could occur as part of a question on how objects move or have their temperature changed, as part of a question about an experiment or investigation, or as stand-alone questions.

Have a go at this exam-style question. Write your answer on paper.

Exam-style question

1 A student is investigating the efficiency of different bouncing balls by dropping them from a height of 2.0 m and measuring the height to which they bounce back up.

The student drops a tennis ball of mass 0.25 kg. After the first bounce it bounces back to a height of 1.40 m.

The gravitational field strength is 10 N/kg.

1.1 Calculate the gravitational potential energy of the tennis ball before it is dropped. **(2 marks)**

1.2 Calculate the maximum speed the tennis ball could be travelling at just before it hits the ground. **(3 marks)**

1.3 Describe the energy transfers during the bounce when the ball is in contact with the ground. **(2 marks)**

1.4 The efficiency of the bounce is less than 1. Calculate the efficiency of the bounce. **(3 marks)**

1.5 Explain what happens to the gas pressure inside the tennis ball as it warms up. Include ideas about kinetic energy, temperature and particles in your answer. **(3 marks)**

Boost your grade

To boost your grade, make sure you know how to use and rearrange the energy equations. Learn the energy equations that are not given on the Physics Equations Sheet. You also need to remember the efficiency equations.

How confident do you feel about each of these **skills**? Colour in the bars.

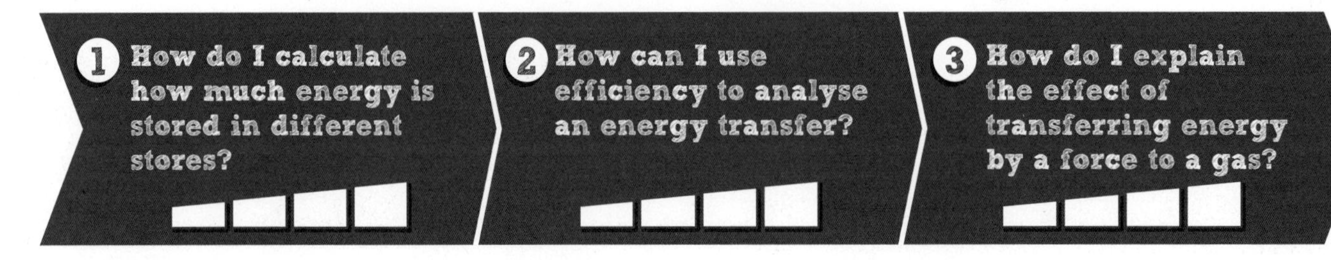

This unit will help you to learn more about the changes in the structure of atomic nuclei that happen during radioactive decay. It will also help you to learn about the patterns that occur during the random decay.

In the exam you will be asked to answer questions such as the one below.

Exam-style question

1 A research scientist measured the count rate produced by a radiation detector for two different radioactive isotopes over a period of time.

The results are shown in **Figure 1**.

1.1 What are isotopes? **(2 marks)**

Same element with different number of neutrons

1.2 Which of the isotopes was most active at the start of the experiment? A **(1 mark)**

1.3 Which of the isotopes had the longest half-life? B **(1 mark)**

Count rate in becquerel (Bq): 0, 300, 600, 900, 1200, 1500, 1800
Time in days: 0, 2, 4, 6, 8, 10, 12, 14, 16
Count rate for isotope A
Count rate for isotope B

Figure 1

1.4 Use the graph to determine the half-life of isotope B. **(1 mark)**

4.6 days

A sample of a different radioactive isotope (iodine-53) is known to decay by beta particle emission.

1.5 Complete the decay equation to show the three missing values. **(3 marks)**

$$^{131}_{53}\mathrm{I} \rightarrow \, ^{131}_{54}\mathrm{Xe} + \, ^{0}_{-1}\mathrm{e}$$

You will already have done some work on atomic structure, radioactivity and radioactive decay. Before starting the **skills boosts**, rate your confidence in each area. Colour in the bars.

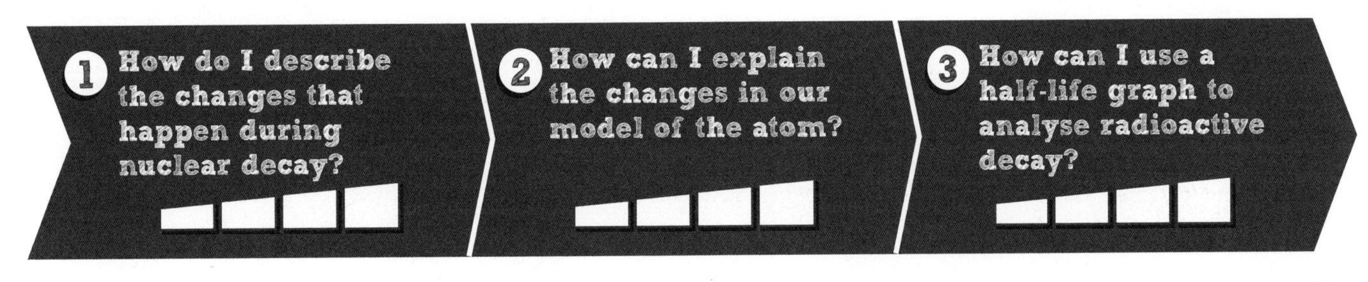

Get started

Atoms are the building blocks of all molecules and materials. Each atom is built from only three types of subatomic particles, **protons**, **neutrons** and **electrons**, in different numbers and arrangements.

1 Cross out (~~cat~~) the incorrect answers in the table to leave the correct descriptions of the properties and locations of protons, neutrons and electrons.

Component	Electric charge	Location
proton	positive / ~~negative~~ / neutral	in the nucleus / ~~different energy levels around the nucleus~~
neutron	~~positive~~ / negative / neutral	in the nucleus / ~~different energy levels around the nucleus~~
electron	positive / negative / ~~neutral~~	~~in the nucleus~~ / different energy levels around the nucleus

Nuclear notation is used to describe a nucleus. This notation contains the element symbol, the atomic number and the mass number in the format shown on the right.

mass number — $^{23}_{11}Na$ — atomic number

2 Complete these statements about the use of nuclear notation. Use words from the box. Each word can be used once or more than once.

atomic number   mass number   protons   neutrons   electrons   nucleus

a The atomic number is the number of ......Protons...... in the nucleus.

b The mass number is the total number of ......Protons...... and ......Neutrons...... in the nucleus.

c The number of neutrons is equal to the .... minus the ....

d In an atom the number of .... is equal to the number of protons in the ....

There are several different variations of carbon atoms. All carbon atoms have 6 protons but the number of neutrons can vary. We say that there are different **isotopes** of carbon. There are also different isotopes of every other element.

3 Here are three different isotopes of chlorine: $^{35}_{17}Cl$ $^{36}_{17}Cl$ $^{37}_{17}Cl$

a Write a sentence to describe what is **the same** for these isotopes. Answer in terms of subatomic particles.

....

b Write a sentence which describes what is **different** about these isotopes. Answer in terms of subatomic particles.

Try to use the terms protons and neutrons, and possibly electrons, in your descriptions.

....

....

4 Use the information in **2** to complete this table for some example isotopes. The table contains only atoms.

Isotope and chemical symbol	Protons	Neutrons	Electrons	Atomic Notation
Carbon-14 C	(from atomic number) 6	Mass no. – atomic no.	(same as protons)	$^{....}_{....}C$
Carbon-12 C	6			

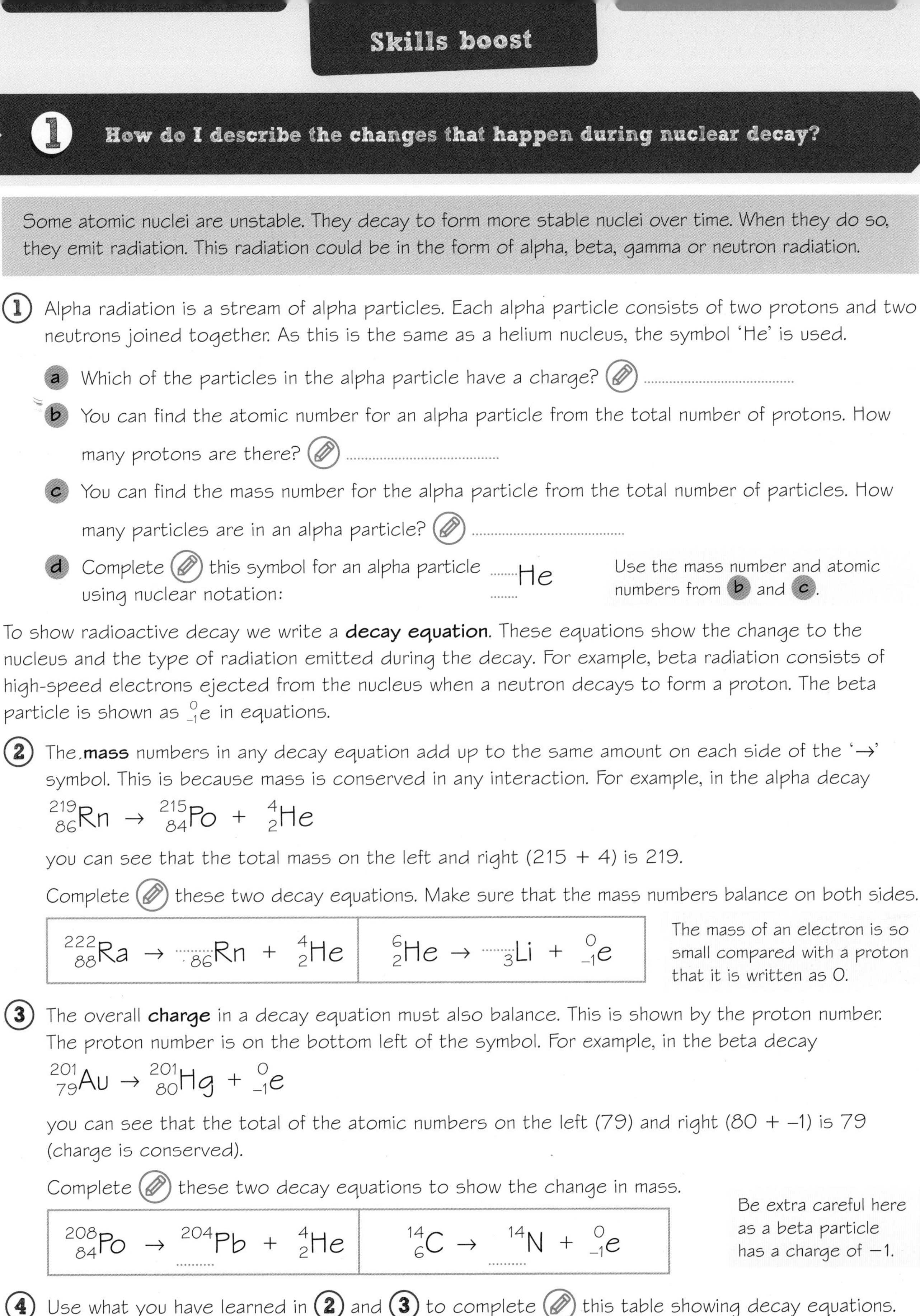

Skills boost

1 How do I describe the changes that happen during nuclear decay?

Some atomic nuclei are unstable. They decay to form more stable nuclei over time. When they do so, they emit radiation. This radiation could be in the form of alpha, beta, gamma or neutron radiation.

1 Alpha radiation is a stream of alpha particles. Each alpha particle consists of two protons and two neutrons joined together. As this is the same as a helium nucleus, the symbol 'He' is used.

a Which of the particles in the alpha particle have a charge? ....

b You can find the atomic number for an alpha particle from the total number of protons. How many protons are there? ....

c You can find the mass number for the alpha particle from the total number of particles. How many particles are in an alpha particle? ....

d Complete this symbol for an alpha particle using nuclear notation: $^{\dots}_{\dots}\text{He}$

Use the mass number and atomic numbers from **b** and **c**.

To show radioactive decay we write a **decay equation**. These equations show the change to the nucleus and the type of radiation emitted during the decay. For example, beta radiation consists of high-speed electrons ejected from the nucleus when a neutron decays to form a proton. The beta particle is shown as $^{0}_{-1}\text{e}$ in equations.

2 The **mass** numbers in any decay equation add up to the same amount on each side of the '→' symbol. This is because mass is conserved in any interaction. For example, in the alpha decay

$$^{219}_{86}\text{Rn} \rightarrow {}^{215}_{84}\text{Po} + {}^{4}_{2}\text{He}$$

you can see that the total mass on the left and right (215 + 4) is 219.

Complete these two decay equations. Make sure that the mass numbers balance on both sides.

$^{222}_{88}\text{Ra} \rightarrow {}^{\dots}_{86}\text{Rn} + {}^{4}_{2}\text{He}$	$^{6}_{2}\text{He} \rightarrow {}^{\dots}_{3}\text{Li} + {}^{0}_{-1}\text{e}$

The mass of an electron is so small compared with a proton that it is written as 0.

3 The overall **charge** in a decay equation must also balance. This is shown by the proton number. The proton number is on the bottom left of the symbol. For example, in the beta decay

$$^{201}_{79}\text{Au} \rightarrow {}^{201}_{80}\text{Hg} + {}^{0}_{-1}\text{e}$$

you can see that the total of the atomic numbers on the left (79) and right (80 + –1) is 79 (charge is conserved).

Complete these two decay equations to show the change in mass.

$^{208}_{84}\text{Po} \rightarrow {}^{204}_{\dots}\text{Pb} + {}^{4}_{2}\text{He}$	$^{14}_{6}\text{C} \rightarrow {}^{14}_{\dots}\text{N} + {}^{0}_{-1}\text{e}$

Be extra careful here as a beta particle has a charge of –1.

4 Use what you have learned in **2** and **3** to complete this table showing decay equations.

Decay type	Equation
alpha	$^{185}_{79}\text{Au} \rightarrow {}^{\dots}_{\dots}\text{Ir} + {}^{4}_{2}\text{He}$
	$^{14}_{6}\text{C} \rightarrow {}^{14}_{7}\text{N} + {}^{\dots}_{\dots}\dots$

Decay type	Equation
	$^{\dots}_{\dots}\text{Pa} \rightarrow {}^{227}_{89}\text{Ac} + {}^{4}_{2}\text{He}$
beta	$^{8}_{3}\text{Li} \rightarrow {}^{\dots}_{\dots}\text{Be} + {}^{0}_{-1}\text{e}$

Skills boost

2 How can I explain the changes in our model of the atom?

The model of an atom has changed over time as new evidence has been discovered through observation and carefully designed experiment.

The diagrams show ideas about the structure of an atom before and after an experiment carried out by Ernest Rutherford and his research team. The experiment involved firing a beam of fast-moving alpha particles at a thin gold foil and detecting which ways the alpha particles were deflected as they reached the foil. The diagrams show the expected and observed paths of alpha particles through a gold atom.

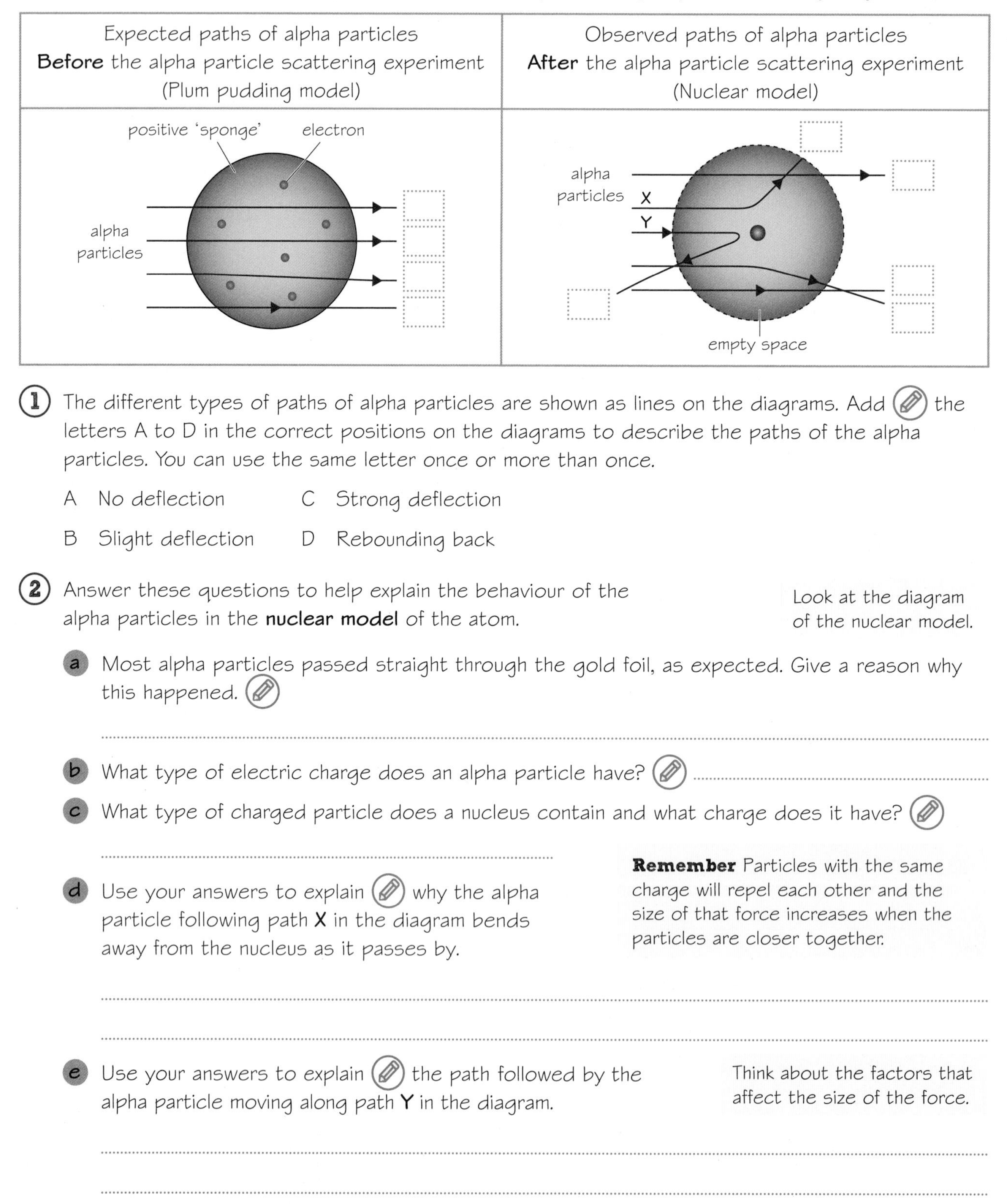

1. The different types of paths of alpha particles are shown as lines on the diagrams. Add the letters A to D in the correct positions on the diagrams to describe the paths of the alpha particles. You can use the same letter once or more than once.

   A No deflection C Strong deflection

   B Slight deflection D Rebounding back

2. Answer these questions to help explain the behaviour of the alpha particles in the **nuclear model** of the atom.

   Look at the diagram of the nuclear model.

   a. Most alpha particles passed straight through the gold foil, as expected. Give a reason why this happened.

   ....

   b. What type of electric charge does an alpha particle have? ....

   c. What type of charged particle does a nucleus contain and what charge does it have?

   ....

   d. Use your answers to explain why the alpha particle following path **X** in the diagram bends away from the nucleus as it passes by.

   Remember Particles with the same charge will repel each other and the size of that force increases when the particles are closer together.

   ....

   ....

   e. Use your answers to explain the path followed by the alpha particle moving along path **Y** in the diagram.

   Think about the factors that affect the size of the force.

   ....

   ....

3 How can I use a half-life graph to analyse radioactive decay?

Radioactive decay is a **random** process. You cannot predict when a particular nucleus will decay. However, when there are a large number of nuclei, you can predict a **pattern** to the decay and work out how many nuclei will be left after a certain time.

The **activity** of a sample is the number of decays that happen per second. Activity is measured in a unit called the becquerel (Bq). This activity falls in a specific pattern giving a **decay curve** shape.

The graph shows the activity of isotope **X** over time.

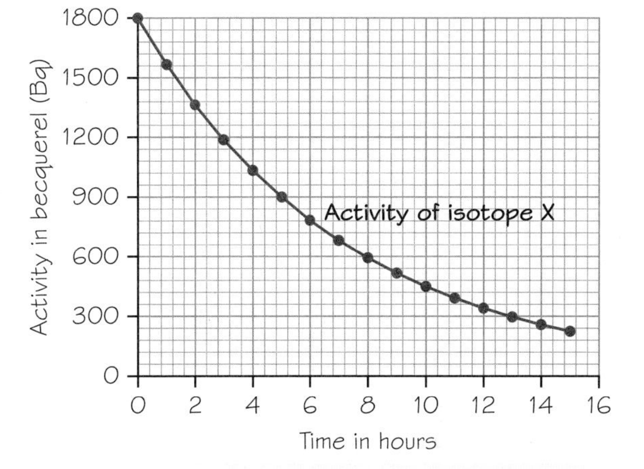

1 Circle Ⓐ the correct answers to the questions about the behaviour of isotope **X**. Draw lines on the graph to help you read the values accurately.

a What is the initial activity of the sample?

180 Bq 1200 Bq 1800 Bq 2000 Bq

b How long does it take for the activity of the sample to halve?

0 hours 2.5 hours 5 hours 10 hours

Use the graph to find the time taken for the activity to fall to half of your answer in **a**.

c How much **longer** does it take for the activity to halve again (1/4 of the original activity)?

0 hours 2.5 hours 5 hours 10 hours

Look for the pattern in the decay.

2 We can use the symbol '→' to represent one half-life passing. What fraction of the original activity will there be after a total time of 20 hours?

$1 \xrightarrow{\text{5 hours}} \frac{1}{2} \xrightarrow{\text{5 hours}} \frac{1}{2} \text{ of } \frac{1}{2} = \frac{1}{4} \xrightarrow{\text{5 hours}} \frac{1}{2} \text{ of } \frac{1}{4} = \frac{1}{8} \xrightarrow{\text{5 hours}} \frac{1}{2} \text{ of } \frac{1}{8} = \frac{1}{16}$

The **half-life** of a radioactive sample is the time it takes for the activity of that sample to fall to half of it original value. The activity (and so the count rate) falls by half every half-life.

3 Complete the table to show the pattern in decay for isotope **X** in **1**.

Time in hours	0		5		10		
Activity in Bq	1800	→		→		→	
Fraction remaining	$\frac{1}{1}$	1st half-life	$\frac{1}{2}$	2nd half-life		3rd half-life	

4 On graph paper, use the data in the table below to plot a graph showing the activity of a second sample over time. Use the same axes as in **1**. Label your graph as isotope **Y**.

Time in hours	0	1	2	3	4	5	6	7	8	9	10	11	12	13	14	15
Activity in Bq	1200	984	808	662	543	446	366	300	246	202	166	136	111	91	75	62

5 Complete this table to show the pattern in decay for isotope **Y**.

Time in hours	0						
Activity in Bq		→		→		→	
Fraction remaining	$\frac{1}{1}$	1st half-life		2nd half-life		3rd half-life	

Sample response

When describing the structure of the nucleus and nuclear decays, remember that:

- The nuclear model of the atom was developed because of the results of an alpha particle scattering experiment which could not be explained by earlier models.
- The changes which happen in radioactive decay should be shown in carefully balanced nuclear decay equations.
- The half-life of an isotope is the time taken for half of the atoms to decay.
- Half-life is often found from a graph of activity over time.

Here are some exam-style questions. Use the student responses to these questions to improve your understanding of radioactive decay.

Exam-style question

1 A radioactive isotope has an initial activity of 16 000 Bq and a half-life of 30 minutes. What will the activity of the sample be after 2 hours? **(2 marks)**

Look at this student's answer.

0 min → 30 min → 60 → 90 → 120 is 5 half-lives. $16000 \times \frac{1}{2} \times \frac{1}{2} \times \frac{1}{2} \times \frac{1}{2} \times \frac{1}{2} = 500$ *Bq.*

The unit 'Bq' stands for **becquerel**. This is the number of decays each second.

1 What does the '→' represent in the student's answer? ....

....

2 The student answer for the half-life is incorrect. The student has counted the numbers (5) and so thinks that five half-lives have passed.

a How many half-lives have really passed? How can you tell? ....

b Rewrite the answer but put in activities (in Bq) instead of times to reach the correct answer.

....

Exam-style question

2 The isotope strontium-90 decays through beta particle emission. Complete the decay equation for the decay of strontium-90. **(3 marks)**

$$^{90}_{38}Sr \rightarrow \,^{\cdots}_{\cdots}Y + \ldots\ldots$$

Look at this student's answer.

$$^{90}_{38}Sr \rightarrow \,^{90}_{37}Y + \,^{0}_{1}e$$

3 The student's nuclear equation is partially correct but a mistake has been made.

a Circle the parts of the equation in the student's answer that are correct.

b Write the correct decay equation.

Be extra careful with beta decays. What happens to the atomic number?

Your turn!

It is now time to use what you have learned to answer the exam-style question from page 49. Remember to read the question thoroughly, looking for information that might help you. Make good use of your knowledge from other areas of physics.

Exam-style question

1 A research scientist measured the count rate produced by a radiation detector for two different radioactive isotopes over a period of time.

The results are shown in **Figure 1**.

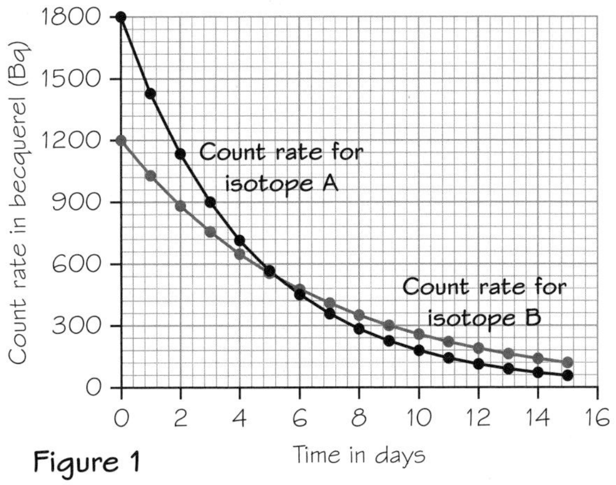

Figure 1

1.1 What are isotopes? **(2 marks)**

....

....

....

....

There are two marks here so make sure you mention one similarity and one difference between the isotopes.

1.2 Which of the isotopes was most active at the start of the experiment? **(1 mark)**

....

Use the graph. Will a more-active source produce a higher or lower count rate?

1.3 Which of the isotopes had the longest half-life? **(1 mark)**

....

You will need to look to see which isotope's count rate halves first. Be careful as the isotopes don't start with the same level of activity.

1.4 Use the graph to determine the half-life of isotope B. **(1 mark)**

....

A sample of a different radioactive isotope (iodine-53) is known to decay by beta particle emission.

1.5 Complete the decay equation to show the three missing values. **(3 marks)**

$$^{131}_{53}I \rightarrow {}^{\dots\dots}_{\dots\dots}Xe + {}^{0}_{\dots\dots}e$$

Watch out for that atomic number again here.

Need more practice?

Questions about atomic structure and radioactive decay could occur as part of a question on radioactive decay and safety, part of a question about an experiment or investigation, or as stand-alone questions.

Have a go at this exam-style question.

Exam-style question

1 In 1910 a research team tested the plum-pudding model of the atom by carrying out an alpha particle scattering experiment. The team fired alpha parties at a thin gold foil and observed the patterns in the scattering of the particles.

Figure 1 shows three alpha particles approaching three nuclei in a layer of gold foil.

1.1 Complete the paths of the three alpha particles shown. **(3 marks)**

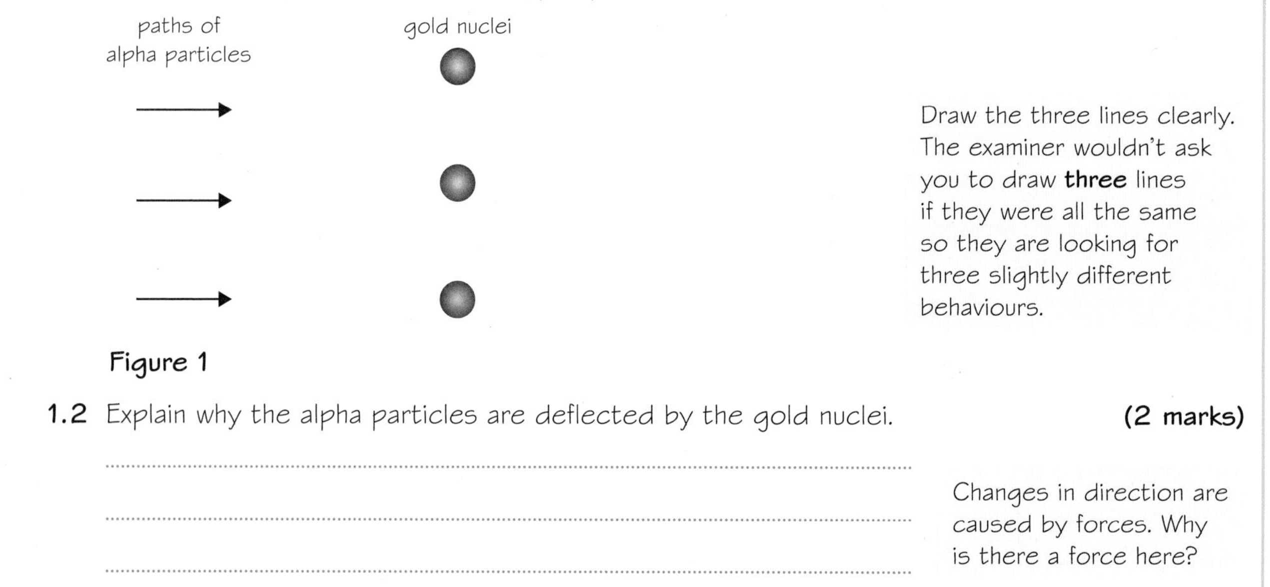

Draw the three lines clearly. The examiner wouldn't ask you to draw **three** lines if they were all the same so they are looking for three slightly different behaviours.

Figure 1

1.2 Explain why the alpha particles are deflected by the gold nuclei. **(2 marks)**

....

....

....

....

Changes in direction are caused by forces. Why is there a force here?

Boost your grade

To boost your grade, make sure that you can describe the four ways in which nuclear radiation may be emitted. You should be able to use equations to explain the changes to nuclei caused by alpha decay and beta decay. You should also make sure that you can explain why scientists changed their model of the nucleus due to the results of the alpha particle scattering experiment and the changes that have happened since the nuclear model was introduced.

How confident do you feel about each of these **skills**? Colour in the bars.

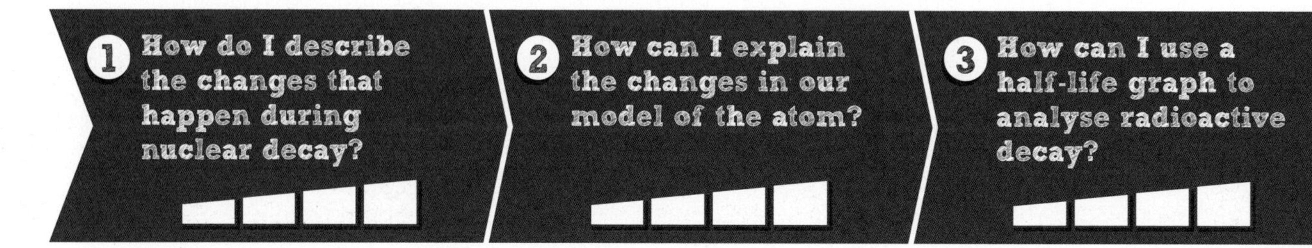

Get started AO2, AO3

8 Equations, calculations and SI units

This unit will help you to answer questions involving physics equations and calculations.

In the exam, you will be asked to answer questions such as the one below.

Exam-style question

1 A student investigated the specific latent heat of ice.

Figure 1 shows the circuit the student used, and how she set up the experiment.

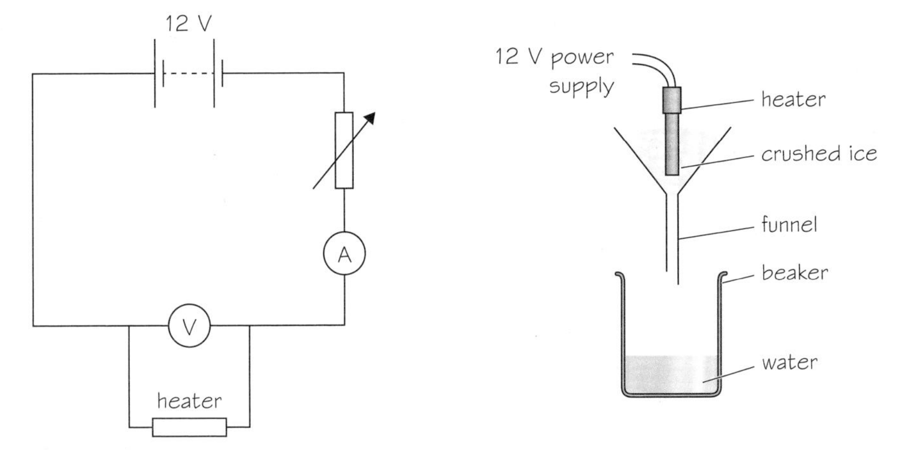

Figure 1

The student set the current to 4.0 A and the potential difference across the heater to 8.0 V using the variable resistor. She ran the experiment for exactly 5 minutes.

1.1 Calculate the power of the electrical heater. **(2 marks)**

1.2 Calculate the energy supplied by the heater. **(2 marks)**

1.3 During the experiment, 34.3 g of ice melted and dripped into the beaker.
Calculate the specific latent heat of fusion of water.
Give your answer to 2 significant figures in J/kg. **(3 marks)**

You will already have done some work on equations, calculations and SI units. Before starting the **skills boosts**, rate your confidence in equations, calculations and SI units. Colour in the bars.

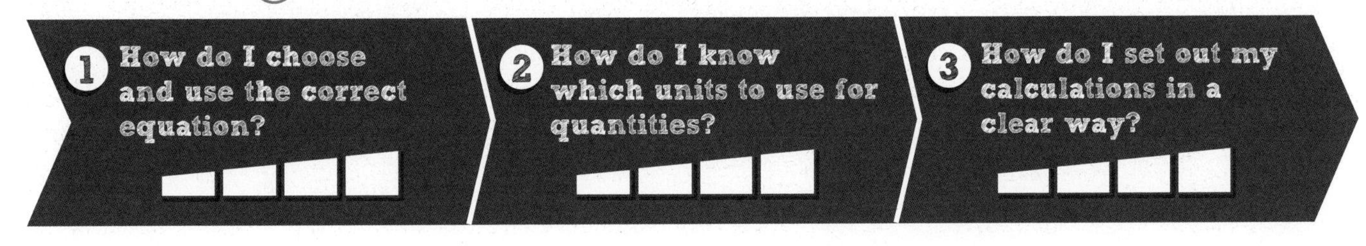

Get started

There are many equations used in physics to describe the relationships between physical quantities. In the exam, you are given some of the equations on the Physics Equations Sheet, but you need to remember others.

Symbols are a way of representing physical quantities and units. For example, the equation momentum = mass × velocity is represented by $p = m\,v$

$m\,v$ means $m \times v$

1 In the equation $p = m\,v$, p and v are lowercase letters. What physical quantities or units would they represent if they were uppercase letters?

P represents .... and ....

V represents ...., ... and ....

The International System of Units is used by all scientists. Units in this system are called SI units.

2 All the physical quantities below have an SI unit. Circle Ⓐ the correct SI unit for each quantity.

Distance or length **centimetre / metre / kilometre**	Time **second / minute / hour**	Energy **joule / kilojoule / megajoule**
Power **watt / kilowatt / megawatt**	Mass **gram / kilogram / tonne**	Current **milliampere / ampere**
Force **newton / kilonewton**	Potential difference **millivolt / volt / kilovolt**	Pressure **pascal / kilopascal**

Make sure you know what the correct SI units are for each physical quantity. Some equations in physics only work when SI units are used. For example, $F = m\,a$ does not give answers in newtons if the mass is in grams – it must be in kilograms!

The base units you will use in physics are: metre, kilogram, second, and ampere. Other units are formed from these base units and are called derived units.

For example, density $= \frac{\text{mass}}{\text{volume}}$, so the unit of density will be the unit of mass divided by the unit of volume, which is kg/m^3.

3 Work out the unit for momentum from the equation: momentum = mass × velocity

....

Sometimes you need to change the subject of an equation. For example, to calculate **mass** using the equation momentum = mass × velocity ($p = m\,v$) you we need to make ***m*** the subject of the equation.

The subject of the equation $p = m\,v$ is ***p***, the letter on its own on one side of the equals sign.

To rearrange the equation to make mass the subject of the equation, divide both sides by ***v***:

$$p = m\,v \qquad \frac{p}{v} = m \times \frac{v}{v}$$

$$\frac{p}{v} = m \times \frac{\cancel{v}}{\cancel{v}}$$

$$\text{giving} \quad \frac{p}{v} = m \text{ or } m = \frac{p}{v}$$

$\frac{v}{v} = 1$

Always do the same to both sides of the equation.

4 Rearrange $F = m\,a$ to make ***a*** the subject of the equation.

How do I choose and use the correct equation?

There are many equations that you need to use and apply. In single science physics there are 23 that you need to remember and 12 that you are given on the Physics Equations Sheet.

Exam-style question

1 A lamp has a current of 1.6 A flowing through it.

Calculate the charge that passes through the lamp in 25 s. **(3 marks)**

First you need to identify the physical quantities. A physical quantity is a physical property that you can measure such as temperature, potential difference or mass.

1 **a** Underline the physical quantities given in the exam-style question above.

The question may not name all the physical quantities; sometimes it will just give the value of the quantity.

b Circle the physical quantity you are being asked to calculate.

c Write the symbols for the three physical quantities.

Remember These are the symbols for the physical quantities rather than the symbols for the units.

....

You now need to choose an equation that contains these three quantities.

d Here are some equations related to electricity.
Circle the equation that contains the three quantities.

$E = QV$	$P = \frac{E}{t}$	$E = IVt$
$Q = It$	$P = VI$	$F = BIl$
$V = IR$	$P = I^2 R$	$V_s I_s = V_p I_p$

e Use the equation you have chosen to calculate the answer. Don't forget to give the unit!

....

....

Remember Learn the unit for each physical quantity. The unit for charge is the coulomb, C.

Some equations can be more challenging than others. For example, the equation linking electrical power with current and resistance has a squared term: (current)^2.

Just like in maths, you use the order of operations (BIDMAS) to work out the right order to do your calculations:
B – brackets, I – indices, D – division, M – multiplication, A – addition, S – subtraction.

2 A mains cable with a resistance of 3 Ω carries a current of 8 A. What is the power dissipated to the surroundings by the cable?
Circle the correct calculation below.

$P = I^2 R$	$P = I^2 R$	$P = I^2 R$
$P = 8^2 \times 3$	$P = 8^2 \times 3$	$P = 8^2 \times 3$
$P = 24$ W	$P = 192$ W	$P = 48$ W

Write in the multiplication signs when you substitute the numbers into the equation.

2 How do I know which units to use for quantities?

Some physical quantities do not have a named unit of their own. For example, the unit for density is derived from the units of mass and volume from which density is calculated.

1 An aluminium cube of side length 0.1 m has a mass of 2.7 kg.

a Calculate the volume of the aluminium cube. Give the unit.

....

b Calculate the density of aluminium. Give the unit.

$$\text{density} = \frac{\text{mass}}{\text{volume}}$$

As the SI unit of mass is kg and the unit of volume is m^3, the SI unit of density is kg/m^3, which means kilograms divided by metres cubed (kilograms per cubic metre).

Units are sometimes given prefixes such as mega, kilo, centi, milli, micro and nano. You need to know how to convert one unit to another. For example, there are 1000 m in 1 km. So, to convert km to m, you need to multiply by 1000. To convert m to km, you need to divide by 1000.

2 Match the prefix to the correct multipliers. One has been done for you.

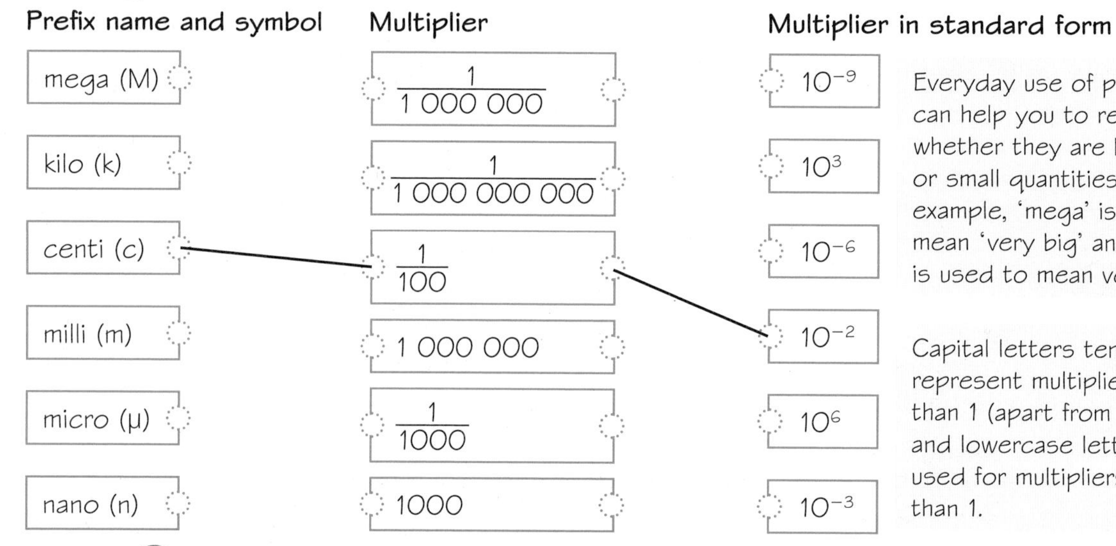

Everyday use of prefixes can help you to remember whether they are big or small quantities. For example, 'mega' is used to mean 'very big' and 'micro' is used to mean very small.

Capital letters tend to represent multipliers bigger than 1 (apart from k for kilo) and lowercase letters are used for multipliers smaller than 1.

3 Convert these quantities.

145 m = ... km	145 m = ... cm	2440 mm = ... m
97.7 MHz = ... Hz	48 mV = ... V	101 300 Pa = ... kPa
2300 W = ... kW		

Questions requiring calculations of speed may sometimes give units of distance in centimetres, metres or kilometres, and units of time in seconds, minutes, hours or even days or years. For example, the average speed of a glacier might be given in cm/day or m/year.

4 Calculate the missing quantities in the table.

Object	Distance	Time	Average speed
car	240 m	40 s	
lizard		15 s	20 cm/s
rocket	480 km		7.5 km/s

For most questions using average speed, you do not need to use SI units. But for questions that use velocity, such as the equations of motion, momentum and kinetic energy, you do need to convert units for distance, time and velocity to SI units.

How do I set out my calculations in a clear way?

It is good practice to structure calculations in a logical way and show your working. Six logical steps are: identify information, choose equation, rearrange, put numbers in, calculate, choose the appropriate number of significant figures and give the unit.

It is important to do this in an exam because you are more likely to get the answer right.

At GCSE, you will usually gain full marks for a correct answer but zero marks for an incorrect answer without any working. You will often be awarded some marks for correct working even if you get the final answer wrong.

Look at this question, and the sample answer below.

A mains electric hoist uses an electric motor to lift a car engine in a workshop.

Calculate the current drawn by the motor when it is working at its full power rating of 0.5 kW. Assume mains electricity is 230 V.

Look at the logical way this calculation has been laid out.

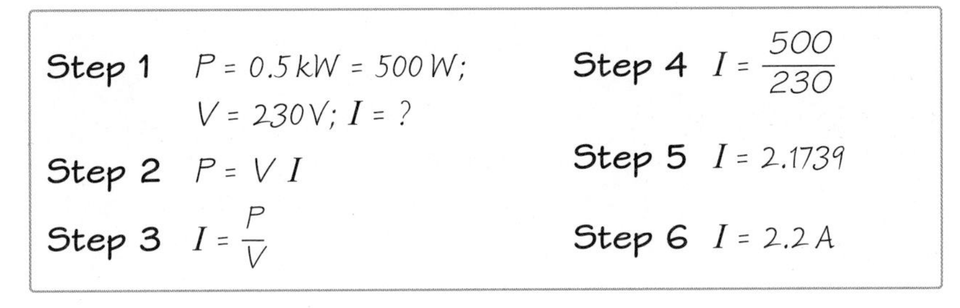

Step 1 $P = 0.5\,kW = 500\,W$; $V = 230\,V$; $I = ?$

Step 2 $P = V\,I$

Step 3 $I = \frac{P}{V}$

Step 4 $I = \frac{500}{230}$

Step 5 $I = 2.1739$

Step 6 $I = 2.2\,A$

Notice how the equals signs in the calculation are all aligned.

To choose the appropriate number of significant figures, look at the largest number of significant figures (sf) given in the values in the question, and give your answer to the same number of significant figures.

1 Write ✎ numbers in the boxes to show the logical order of the steps. One has been done for you.

Choose the right equation and write it down.	2
Calculate the answer and give the unit.	
Identify the physical quantities, making sure the units are SI units if needed.	
Choose the number of significant figures and give the unit.	
Put the numbers in.	
Rearrange the equation if needed.	

2 A small electrical immersion heater supplies 6.25 kJ of energy to an insulated copper block of mass 1.00 kg. The block is initially at 20 °C and the maximum temperature it reaches is 36 °C.

Calculate the specific heat capacity of copper using:

change in thermal energy = mass × specific heat capacity × temperature change

Fill in ✎ the table to complete the calculation.

Step	Calculation
Identify information	change in thermal energy (ΔE) = 6.25 kJ = 6250 J change in temperature ($\Delta\theta$) = ... specific heat capacity (c) = ?
Choose equation	$\Delta E = m\,c\,\Delta\theta$
Re-arrange	$c = \frac{\square}{\square}$
Put in values	$c = \frac{6250}{1.00 \times 16}$
Calculate	c = ...
Choose number of sf and give units	c = ... unit ....

Sample response

Remember that many calculation questions are worth several marks. If you have a logical approach to answering these questions (like the six-step approach), you are more likely to gain full marks.

Exam-style question

1 A student investigated the specific heat capacity of brass.

He used a brass cylinder with holes drilled for an immersion heater and a thermometer.

The brass cylinder had a mass of 0.50 kg.

The immersion heater had a power output of 50 W.

The student heated the brass cylinder for 3 minutes.

Figure 1 shows the equipment the student used.

thermometer

power supply

immersion heater

brass cylinder

Figure 1

1.1 Calculate the energy transferred by the immersion heater in 3 minutes. **(3 marks)**

1.2 The maximum temperature rise of the brass cylinder was 45 °C.

Calculate the specific heat capacity of brass. **(3 marks)**

Look at the sample student answer on the right.

1.1 $P = \frac{E}{t}$

$E = P t$

$E = 50 \times 3$

$E = 150$

1.2 $\Delta E = m\, c\, \Delta\theta$

$c = \frac{\Delta E}{m \Delta\theta}$

$c = \frac{150}{0.5 \times 45}$

$c = 6.666667$

$c = 6.7 \text{ J/kg °C}$

1 **a** The student has given the wrong answer to **1.1**. What two mistakes has the student made?

....

....

b Calculate the correct value of E.

c The student gained full marks for his answer to **1.2**. Explain why he gained full marks even though he used an incorrect value for the energy transferred by the heater.

....

....

d Calculate the correct value for the specific heat capacity of brass laying out your calculation in a logical way.

Your turn!

It is now time to use what you have learned to answer the exam-style question from page 57. Remember to read the question thoroughly, looking for information that may help you. Make good use of your knowledge from other areas of physics.

Exam-style question

1 A student investigated the specific latent heat of ice.

Figure 1 shows the circuit the student used, and how she set up the experiment.

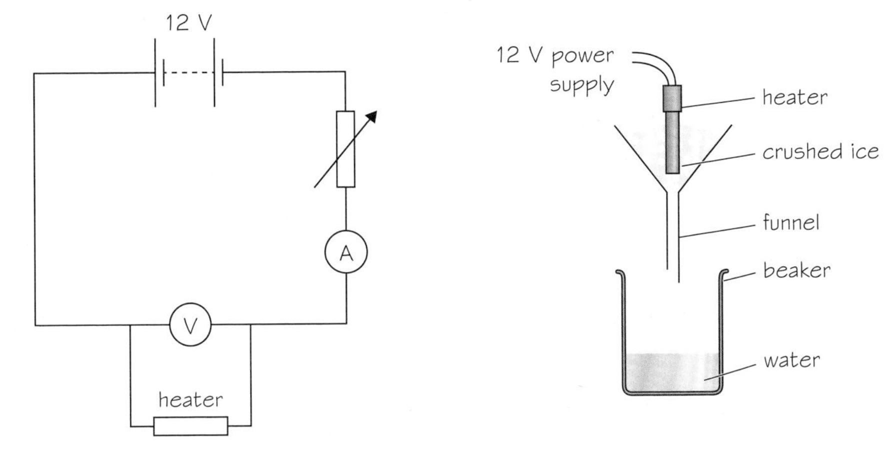

Figure 1

The student set the current to 4.0 A and the potential difference across the heater to 8.0 V using the variable resistor. She ran the experiment for exactly 5 minutes.

1.1 Calculate the power of the electrical heater. **(2 marks)**

Use an equation from the Physics Equations Sheet.

1.2 Calculate the energy supplied by the heater. **(2 marks)**

1.3 During the experiment, 34.3 g of ice melted and dripped into the beaker.

Calculate the specific latent heat of fusion of water.

Give your answer to 2 significant figures in J/kg. **(3 marks)**

Remember Follow the six logical steps: identify information, choose equation, rearrange, put numbers in, calculate, check number of significant figures and give unit.

Need more practice?

Questions about equations, calculations and SI units could occur as part of a question in many topics, or as part of a question about an experiment or investigation.

Have a go at this exam-style question.

Exam-style question

1 A 50 km section of National Grid power line has a resistance of 2.7 Ω and carries a current of 5000 A.

1.1 Give the equation that links resistance and current with electrical power. **(1 mark)**

....

1.2 Calculate the power loss in the power line.

Give your answer in megawatts, MW. **(3 marks)**

....

....

....

....

1.3 A second power line has the same resistance as the first power line and carries half the current.
Describe how this affects the power loss in the second power line.
Justify your descriptions with any relevant calculations. **(4 marks)**

....

....

....

....

....

....

To get high marks, back up your answer to **1.3** with a second calculation.

Boost your grade

To improve your grade, make sure you learn the 23 equations you need to remember. Make sure too that you learn the symbols for the physical quantities in the equations and their SI units.

Practise using the physics equations, making sure you use the six logical steps to lay out your calculations. Remember to use good maths skills such as order of operations (BIDMAS) in calculations, and to use inverse operations when changing the subject of an equation, and to do the same to both sides.
Also remember to look at the number of significant figures the values are given to in the question. Make sure your answer has the same number of significant figures.

How confident do you feel about each of these **skills**? Colour in the bars.

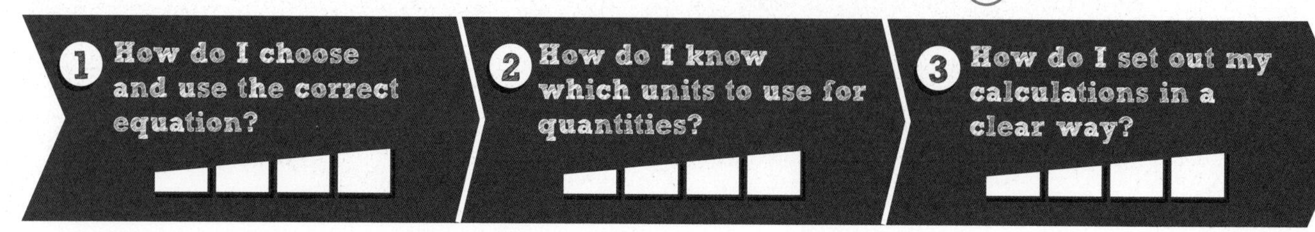

Get started AO1

9 Answering extended response questions

This unit will help you learn more about how to answer exam questions which involve more-extended writing. These questions usually award six marks and require either the detailed understanding of a process or the drawing together of ideas from several areas of physics. Extended response questions often appear as the final part of a multi-part question. The earlier parts of the question can provide information which will help in understanding the question.

In the exam, you will be asked to answer questions such as the one below.

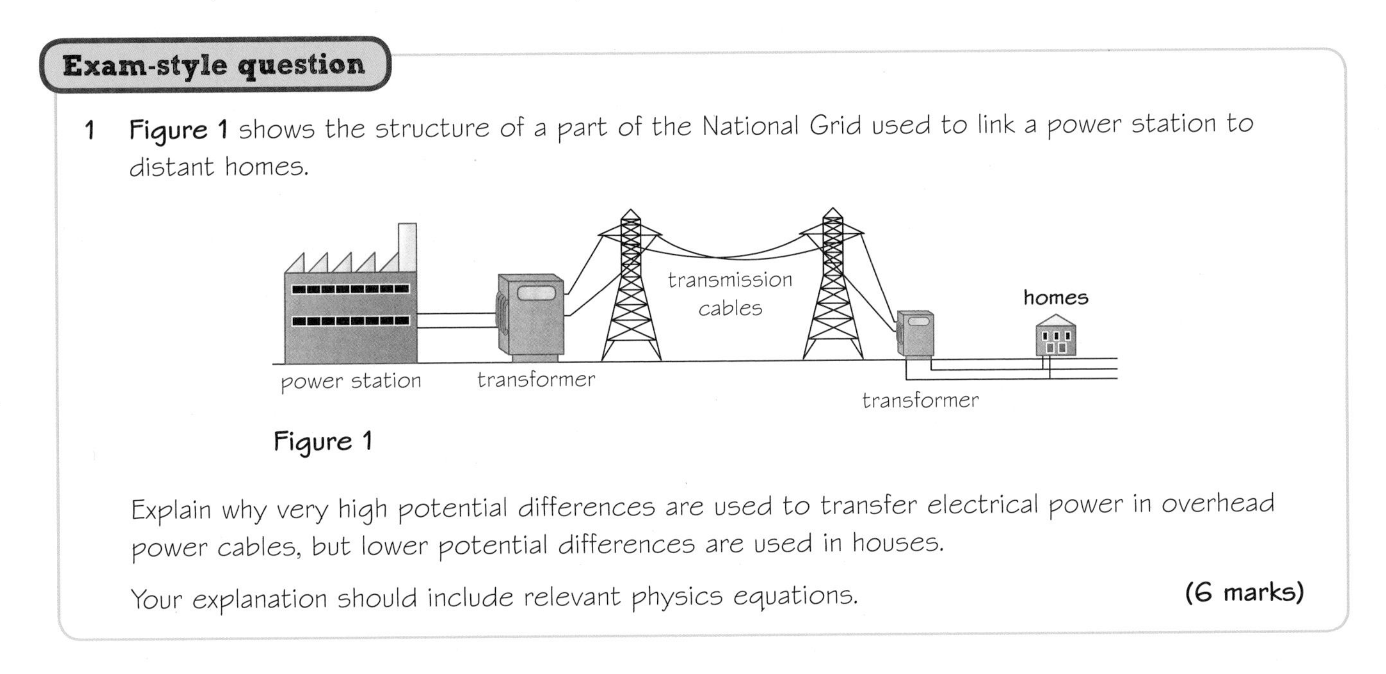

Exam-style question

1 **Figure 1** shows the structure of a part of the National Grid used to link a power station to distant homes.

Figure 1

Explain why very high potential differences are used to transfer electrical power in overhead power cables, but lower potential differences are used in houses.

Your explanation should include relevant physics equations. **(6 marks)**

You will already have done some work about how to answer extended response questions. Before starting the **skills boosts**, rate your confidence in each area. Colour in the bars.

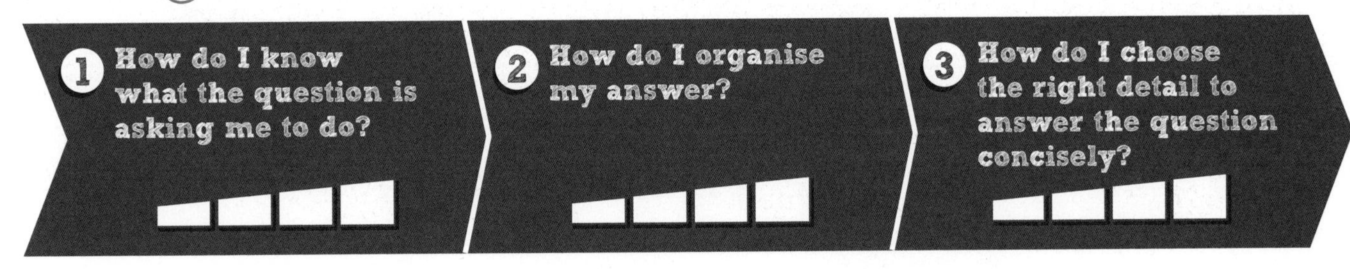

Get started

Extended response questions use different **command words**. These words tell you how to answer the question. For example, whether to write a one-word answer or to provide a detailed explanation.

1 The boxes below show four question parts.

a Underline the command words.

b Draw lines to connect each question part to the correct meaning of the command word and then to the correct style of answer.

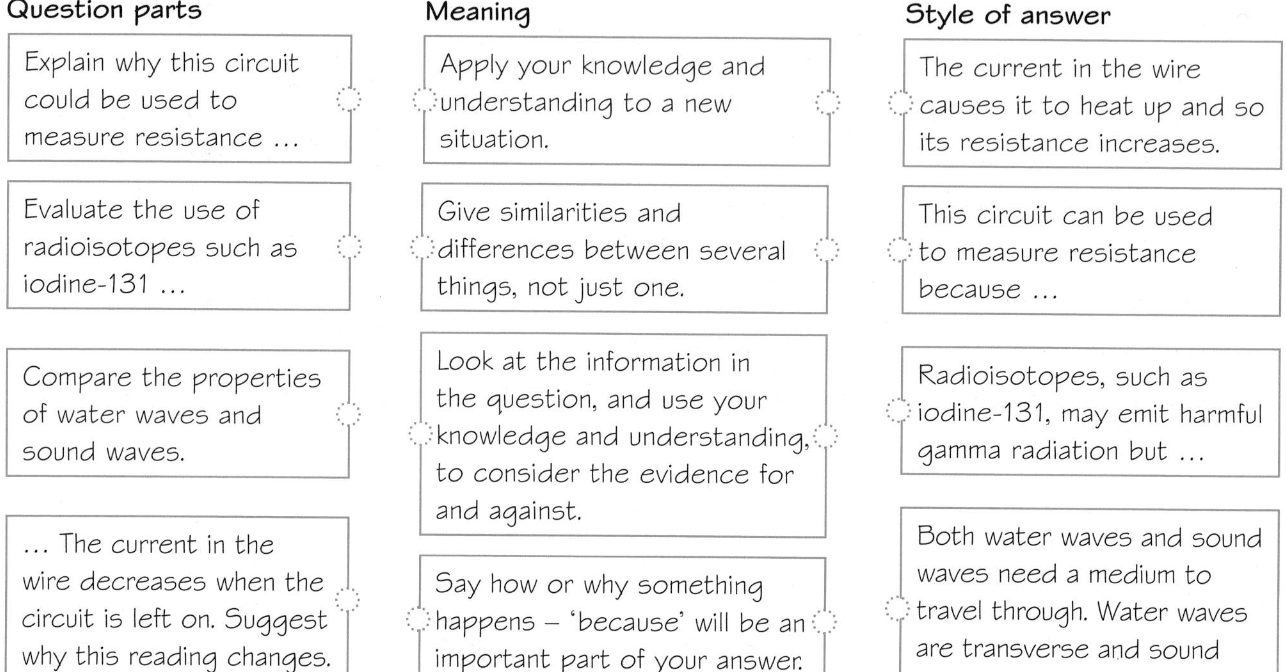

Extended response questions require **planning** as the answer has to contain the correct scientific information in a clear and logical order.

2 A student plans an experiment to find out if different coloured surfaces affect the rate at which they emit infrared radiation and cool down.

Her plan, below, contains the scientific information for a full answer, but it is not in a clear order.

Number the stages of the method so that they are in a logical order. The first one has been done for you.

Stage	Description
1	*Three boiling tubes, one painted black, one white and one silver, are placed in a test-tube rack.*
	Using a stopwatch, measure the temperature of the water in the tube each minute for 10 minutes and record this data in a table.
	Repeat the process with the remaining two boiling tubes.
	Place the thermometer into the black boiling tube. Pour 20 cm³ of boiling water into the boiling tube.
	Compare the cooling of the tubes – the one which has cooled most during the 10 minutes is the best emitter of infrared.
	Use the thermometer to measure the temperature of the water in the boiling tube and wait until the temperature reaches 80 °C.

When you write a method for an experiment, the information given needs to be clear enough for someone else to follow the method and get the results that you expected.

How do I know what the question is asking me to do?

You need to be able to identify the command word and to know what it means.

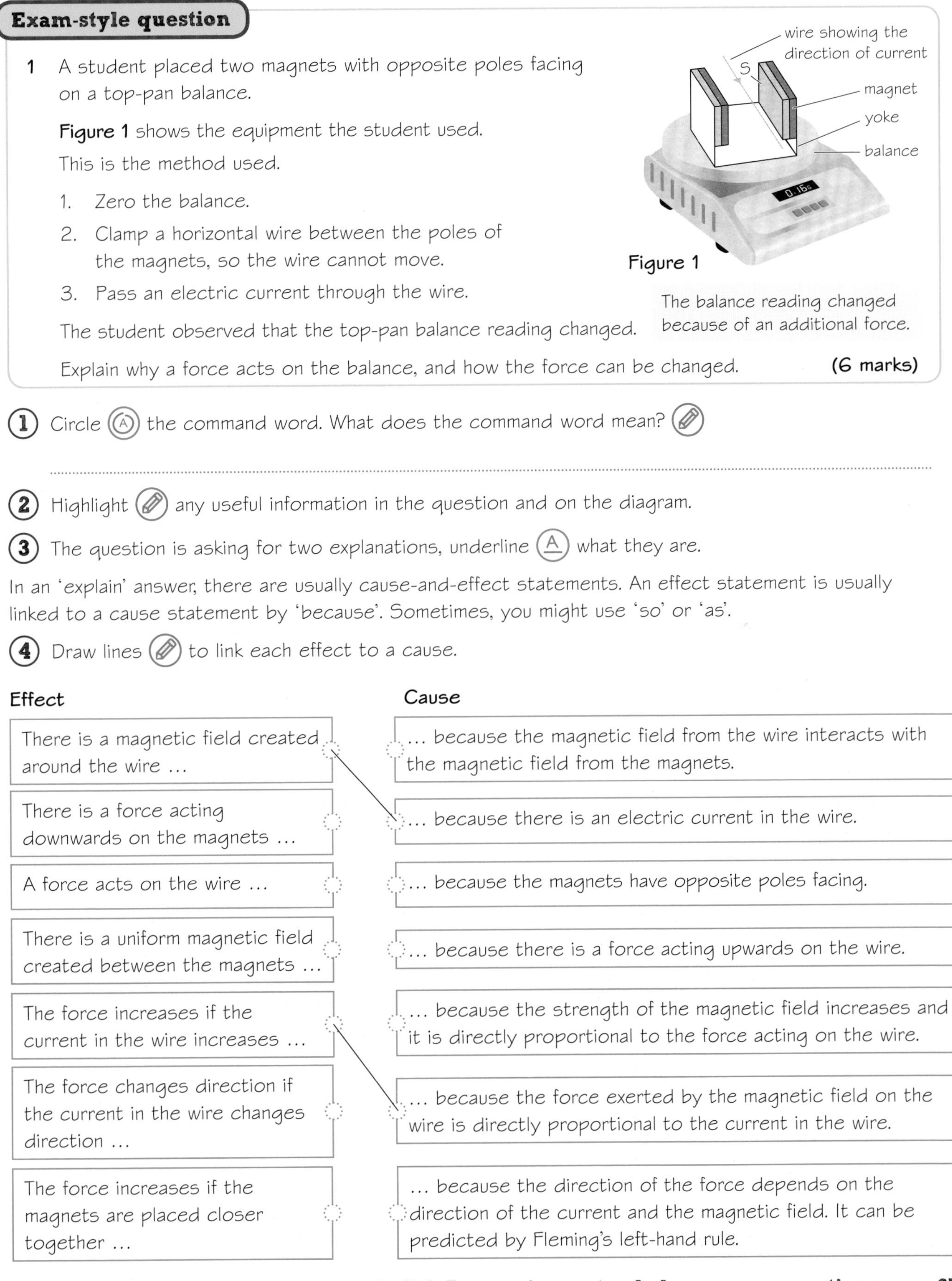

Exam-style question

1 A student placed two magnets with opposite poles facing on a top-pan balance.

Figure 1 shows the equipment the student used.

This is the method used.

1. Zero the balance.
2. Clamp a horizontal wire between the poles of the magnets, so the wire cannot move.
3. Pass an electric current through the wire.

The student observed that the top-pan balance reading changed.

Figure 1

The balance reading changed because of an additional force.

Explain why a force acts on the balance, and how the force can be changed. **(6 marks)**

1 Circle the command word. What does the command word mean?

....

2 Highlight any useful information in the question and on the diagram.

3 The question is asking for two explanations, underline what they are.

In an 'explain' answer, there are usually cause-and-effect statements. An effect statement is usually linked to a cause statement by 'because'. Sometimes, you might use 'so' or 'as'.

4 Draw lines to link each effect to a cause.

Effect

- There is a magnetic field created around the wire …
- There is a force acting downwards on the magnets …
- A force acts on the wire …
- There is a uniform magnetic field created between the magnets …
- The force increases if the current in the wire increases …
- The force changes direction if the current in the wire changes direction …
- The force increases if the magnets are placed closer together …

Cause

- … because the magnetic field from the wire interacts with the magnetic field from the magnets.
- … because there is an electric current in the wire.
- … because the magnets have opposite poles facing.
- … because there is a force acting upwards on the wire.
- … because the strength of the magnetic field increases and it is directly proportional to the force acting on the wire.
- … because the force exerted by the magnetic field on the wire is directly proportional to the current in the wire.
- … because the direction of the force depends on the direction of the current and the magnetic field. It can be predicted by Fleming's left-hand rule.

2 How do I organise my answer?

Make sure you are clear about what the question is asking you to do from the command word. You also need to think about which physics ideas are useful for the answer.

Consider this question again.

Exam-style question

1 A student placed two magnets with opposite poles facing on a top-pan balance.

Figure 1 shows the equipment the student used.

This is the method used.

1. Zero the balance.
2. Clamp a horizontal wire between the poles of the magnets, so the wire cannot move.
3. Pass an electric current through the wire.

The student observed that the top-pan balance reading changed.

Explain why a force acts on the balance, and how the force can be changed. **(6 marks)**

wire showing the direction of current
S
magnet
yoke
balance

Figure 1

Before starting to write your answer, it is a good idea to think about the physics topic the question relates to.

1 Circle (A) the physics topic that is covered in the question.

Energy stores and transfers	The motor effect
Resistance and Ohm's law	Newton's Second Law

It is important to recognise the areas of physics you will use, but you should not write everything you know about that topic.

2 Here are some notes you might use in a concise plan. Number them in a logical order.

Fleming's left-hand rule		*current on wire – magnetic field*	1
force increases if magnets closer together, so strength of magnetic field increases		*force increases if current increases*	
F = B I l		*extra downward force on balance*	
upwards force on wire (Fleming's left-hand rule)		*direction of current changes direction of force*	
magnetic field between magnets		*two magnetic fields interacting – force*	
downwards force on magnets (resultant forces)	4		

Look carefully at the question to make sure you understand what you need to do. This question has two parts, so first you need to explain the force on the balance and then how to change the force.

Make sure you include any appropriate terms and/or equations.

3 Using the cause-and-effect statements from page 67 and your plan, write your answer to the exam-style question on paper.

3 How do I choose the right detail to answer the question concisely?

In your answer, write one or two sentences about each idea.

Exam-style question

1 A ray of light is shone at a rectangular glass block at a shallow angle as shown in **Figure 1**.

Explain what will happen to a ray of light at the air–glass boundary, inside the glass block, and at the glass–air boundary in terms of reflection, refraction, absorption and transmission of light. **(6 marks)**

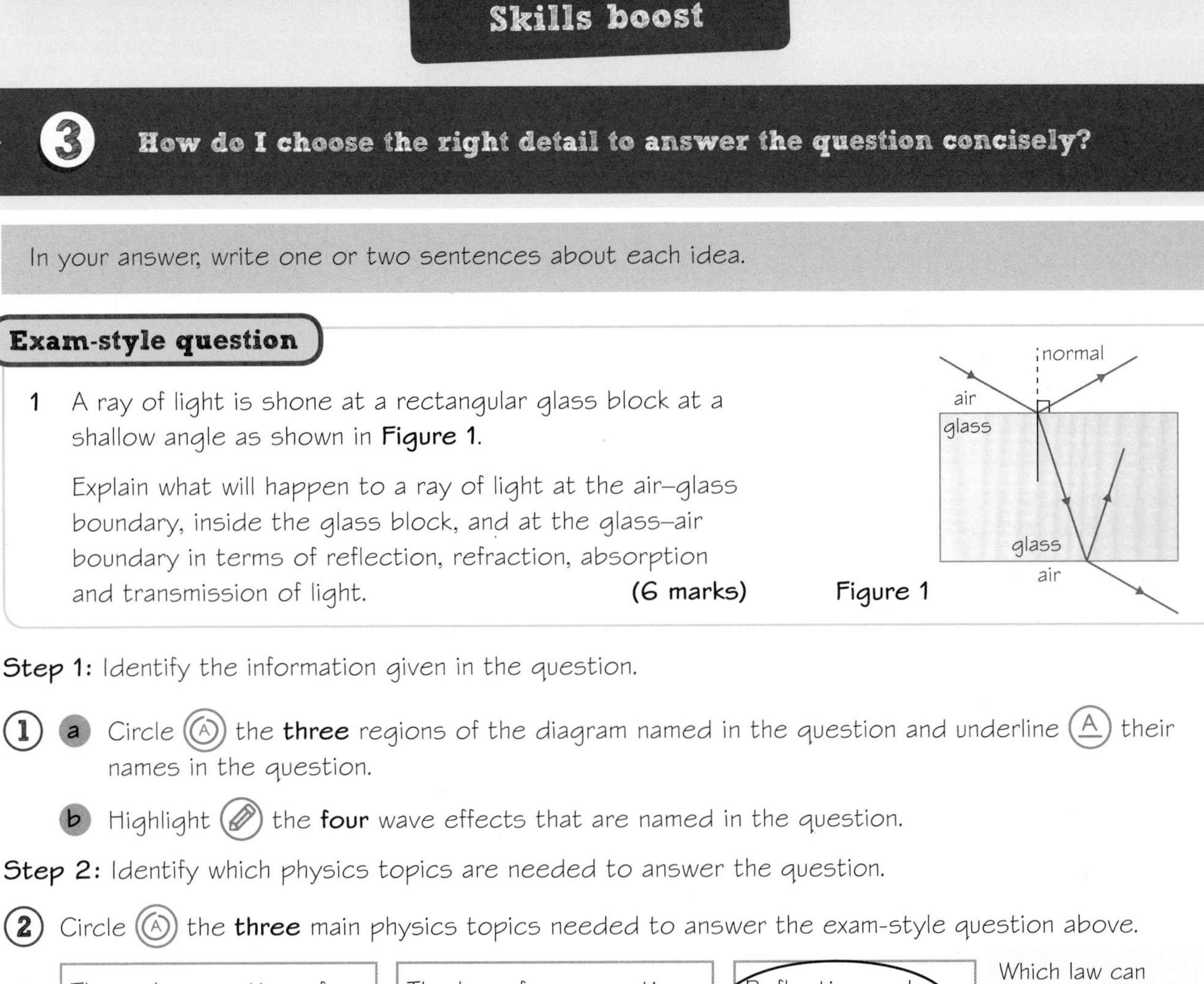

Figure 1

Step 1: Identify the information given in the question.

1 a Circle the **three** regions of the diagram named in the question and underline their names in the question.

b Highlight the **four** wave effects that are named in the question.

Step 2: Identify which physics topics are needed to answer the question.

2 Circle the **three** main physics topics needed to answer the exam-style question above.

Thermal properties of materials	The law of conservation of energy	Reflection and refraction
Absorption and emission of wave energy	Velocity and acceleration	Particle theory

Which law can explain what happens to energy, in this case as a light ray? Can the energy be lost or destroyed?

Step 3: To answer the question accurately, break the question into parts and think about what is happening in each part. For this question think about what happens to the light in each region.

3 a Complete the table using the diagram to help you. One box has been completed for you.

Region	Reflection	Refraction	Absorption	Transmission
air–glass boundary	some light			
inside the glass block				
glass–air boundary				

b Now write about what is happening in each region by completing the sentences below. One has been done for you. Remember to think about cause and effect using 'because'.

As the ray of light reaches the air–glass boundary, some light is reflected but the remaining light is refracted at the surface and enters the block **because** all energy from the light ray must be either refracted or reflected.

Remember Light is a form of energy.

As the ray of light passes through the glass block ....

....

As the ray of light reaches the glass–air boundary ....

....

Sample response

To answer an extended response question, you need to:

- analyse the question to decide what the question is asking for
- identify the scientific ideas that are relevant
- put the ideas in a logical order and make connections between them
- give the right amount of detail in your answer without writing too much.

Exam-style question

1 A frame holding two permanent magnets is placed onto a top-pan balance as shown in **Figure 1**.

The top-pan balance shows the weight of the magnets and frame.

A copper rod is positioned so that it is held in a fixed position between the magnets.

An electric current is passed through the rod in the direction shown.

Predict whether the reading on the scale of the top-pan balance decreases or increases when there is a current in the rod.

Explain your prediction. **(6 marks)**

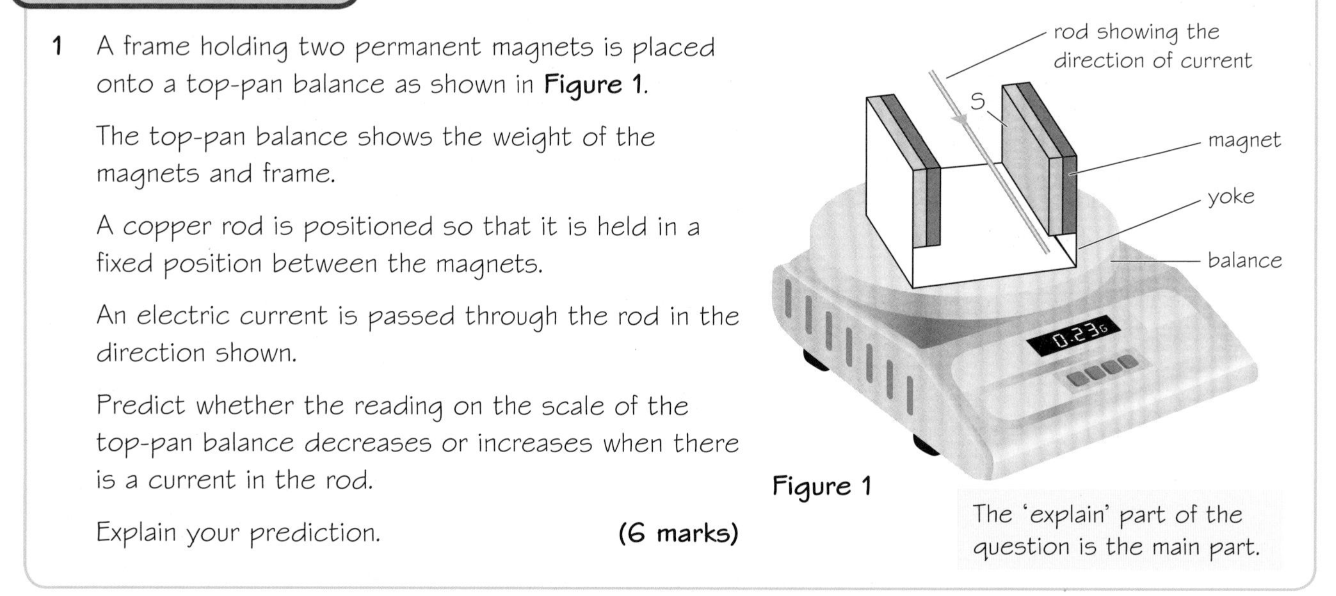

Figure 1

The 'explain' part of the question is the main part.

Look at this student's answer to the exam-style question.

I think it will decrease. Copper is a good conductor but is not normally magnetic but the current in the rod is making it an electromagnet as it passes through. This magnetic rod is then affecting the magnets because there is a force whenever two magnets are placed near to each other because of their magnetic fields. The magnets act to repel each other so the rod is pushed away from the permanent magnets. When magnets put a force on the rod it is the same force that pushes back on the magnets and then the top-pan balance. If you make the current in the rod larger then the reading on the balance will change more because there will be an even larger force because of the equation $F = B I l$.

1 Cross out (~~cat~~) any irrelevant information the student included.

The student's answer needs to be more precise and use the correct terms. There is also information missing.

2 **a** Highlight the sentences where the student attempts to explain why there is a force acting on the wire.

b Underline (A) the sentence where the student attempts to describe the pair of forces acting between the magnets and the rod.

3 Use your knowledge to plan and rewrite the student answer so that it gains full marks. Write your answer on a separate sheet of paper.

Your turn!

It is now time to use what you have learned to answer the exam-style question from page 65. Remember to read the question thoroughly, looking for information that may help you. Make good use of your knowledge from other areas of physics.

Exam-style question

1 **Figure 1** shows the structure of a part of the National Grid used to link a power station to distant homes.

power station | transformer | transmission cables | transformer | homes

Figure 1

Explain why very high potential differences are used to transfer electrical power in overhead power cables, but lower potential differences are used in houses.

Your explanation should include relevant physics equations. **(6 marks)**

1 Circle Ⓐ the command word.

What does the command word mean? Keep that as the focus of your answer.

2 To make sure you fully understand the question, underline Ⓐ the information in the question you have to explain.

You need to identify any relevant equations you might need. The equation that shows electrical power transmitted by a wire or cable is: power = potential difference × current, $P = V\ I$.

3 Explain what $P = V\ I$ tells you about how to transmit a large electrical power.

....

....

The equation that links the current and resistance of a wire to the power it wastes due to heating is: power = current^2 × resistance, $P = I^2\ R$.

4 Explain what this equation tells you about how to reduce the power wasted due to electrical heating.

....

5 Compare your answers to **3** and **4** and explain why high voltages are used to transmit power in long cables.

....

....

6 Write a plan for your answer on paper.

7 Now write your answer on paper. Use the information in your answers to **1** to **6** to help.

Need more practice?

In an exam, extended response questions could occur as:
- simple standalone questions
- part of a question about any topic you have studied
- part of a question about an experimental procedure.

Have a go at this exam-style question. If you need more space to write your answer, continue on paper.

Exam-style question

1 There are many different types of wave including sound waves, water waves and electromagnetic waves.

Some wave properties are common to all waves. Other wave properties are unique to certain types of wave.

Compare the properties of electromagnetic waves and sound waves. **(6 marks)**

Circle the command word in the question, then underline the things you need to compare. This will help you to form a plan.

When you compare, you need to give similarities and differences between two or more things.

Think about the similarities of the two types of wave – these are the common properties of waves. List three or four if you can.

Think about the differences between the two types of wave. Again, list three or four if you can.

Boost your grade

To boost your grade, make sure that you describe all the relevant points clearly using the information provided and your physics knowledge. You should annotate diagrams to help you understand them and plan your answer so that you make the required number of scientific points.

How confident do you feel about each of these **skills**? Colour in the bars.

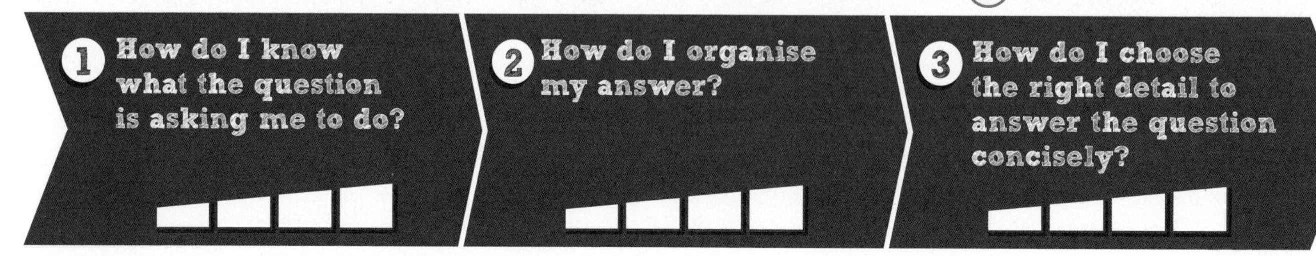

Answers

Unit 1

Page 2

1.

effort	a simple machine consisting of a bar rotating about a pivot
lever	at 90° <u>or</u> normal to <u>or</u> at right angles
load	the force exerted on a lever to lift a load
moment	the axis around which a lever can rotate
perpendicular	the turning effect of a force
pivot	the force exerted on one end of a lever

2. newtons / N
   metres / m
   newton metres / Nm
3. pliers, crow bar, hammer, tweezers, door, nutcrackers, fishing rod, limbs, spade, broom, crane, spanner, etc
4. a clockwise   b type 1
   c small   d small

Page 3

1. a, b and c

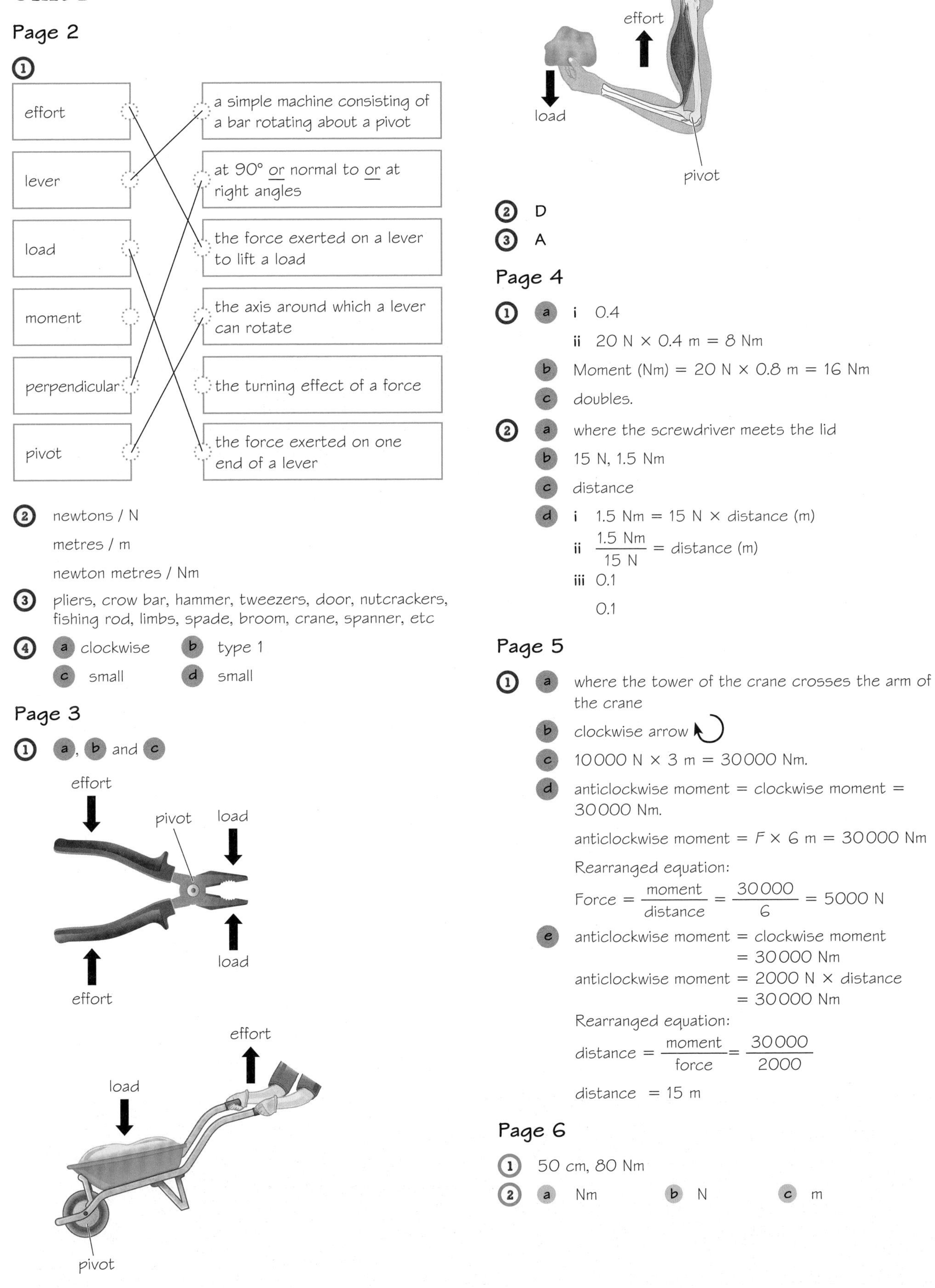

2. D
3. A

Page 4

1. a i 0.4
   ii 20 N × 0.4 m = 8 Nm
   b Moment (Nm) = 20 N × 0.8 m = 16 Nm
   c doubles.
2. a where the screwdriver meets the lid
   b 15 N, 1.5 Nm
   c distance
   d i 1.5 Nm = 15 N × distance (m)
   ii $\frac{1.5 \text{ Nm}}{15 \text{ N}}$ = distance (m)
   iii 0.1
   0.1

Page 5

1. a where the tower of the crane crosses the arm of the crane
   b clockwise arrow ↻
   c 10000 N × 3 m = 30000 Nm.
   d anticlockwise moment = clockwise moment = 30000 Nm.
   anticlockwise moment = F × 6 m = 30000 Nm
   Rearranged equation:
   $\text{Force} = \frac{\text{moment}}{\text{distance}} = \frac{30000}{6} = 5000 \text{ N}$
   e anticlockwise moment = clockwise moment = 30000 Nm
   anticlockwise moment = 2000 N × distance = 30000 Nm
   Rearranged equation:
   $\text{distance} = \frac{\text{moment}}{\text{force}} = \frac{30000}{2000}$
   distance = 15 m

Page 6

1. 50 cm, 80 Nm
2. a Nm   b N   c m

3. 50 cm; not converted 50 cm to 0.5 m
4. moment (Nm) = force (N) × distance (m)

   80 Nm = force (N) × 0.5 m

   force (N) = $\frac{80 \text{ Nm}}{0.5 \text{ m}}$ = 160 N
5. force
6. perpendicular distance of force from pivot
7. larger same distance
8. **a** 2

   b By stating that both the force and the distance will increase, the moment can only increase.

Page 7

1. **1.1** i moment of a force = force × distance **(1)**

   ii 350 N, 1.5 m

   iii moment = 350 N × 1.5 m **(1)**

   iv moment = 525 (1) unit: Nm **(1)**

   1.2 i 525 Nm

   ii total clockwise moment = total anticlockwise moment

   iii 500 N

   iv distance = $\frac{\text{moment}}{\text{force}}$ **(1)**

   v 525 = 500 × d **(1)**

   vi distance = 1.05 m

Page 8

1. **1.1** anticlockwise moment (Nm)
   = force (N) × perpendicular distance (m) **(1)**

   = 650N × 2m **(1)**

   = 1300 Nm **(1)**

   anticlockwise moment = 1300 Nm

   1.2 clockwise moment = 500 N × 2 m = 1000 Nm **(1)**

   1300 – 1000 = 300 Nm to be provided by 200 N person **(1)**

   anticlockwise moment = clockwise moment

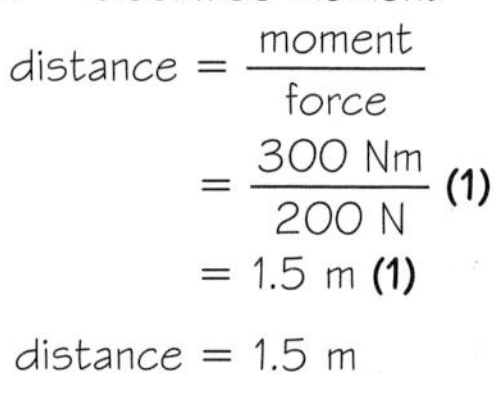

   distance = $\frac{\text{moment}}{\text{force}}$

   = $\frac{300 \text{ Nm}}{200 \text{ N}}$ **(1)**

   = 1.5 m **(1)**

   distance = 1.5 m

Unit 2

Page 10

1. Arrow pointing from source (light) to detector (eye). It should be drawn in the middle of the ray and not at the ends.
2. source; detector; energy
3. top arrow: reflect; middle arrow: absorb; bottom arrow: transmit
4. **a** and **b**

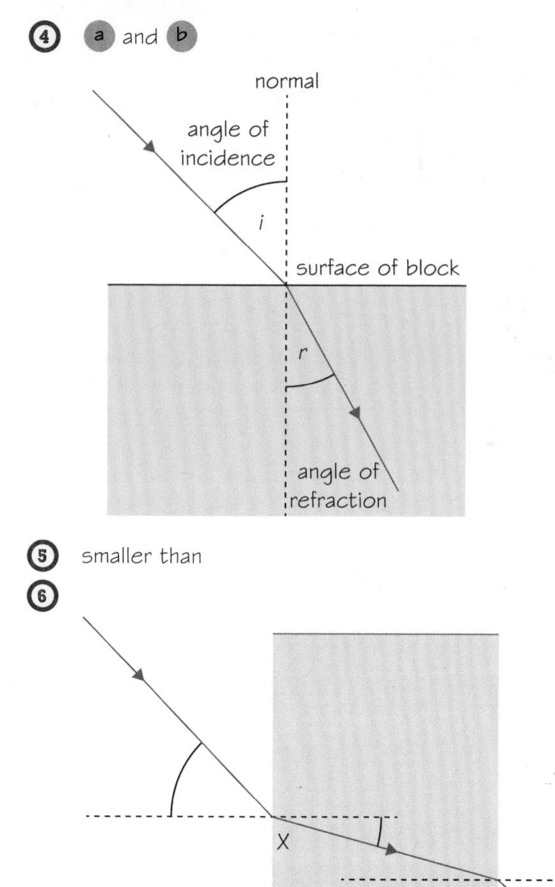

5. smaller than
6.

X

Page 11

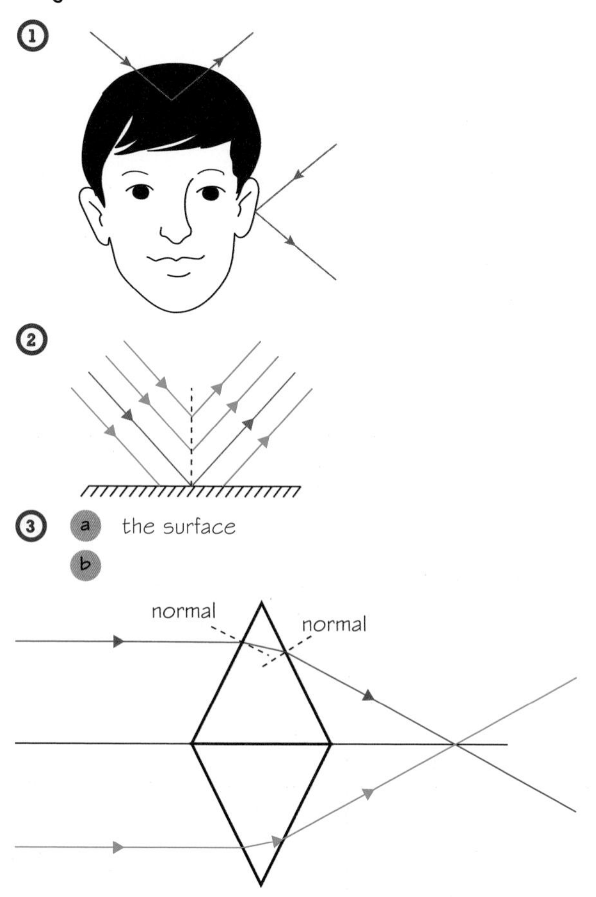

1.
2.
3. **a** the surface

   b

4

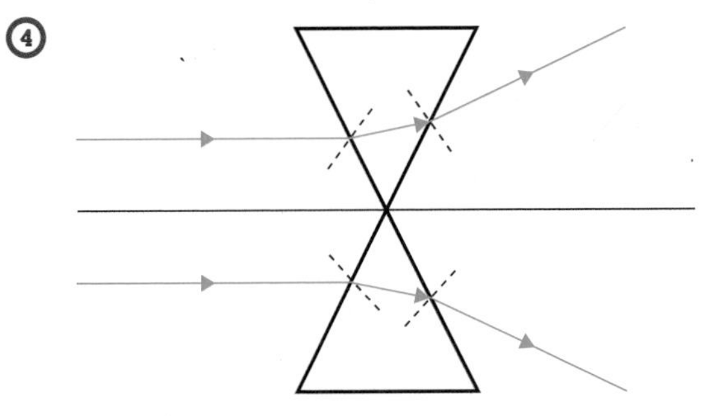

Page 12

1 a and b

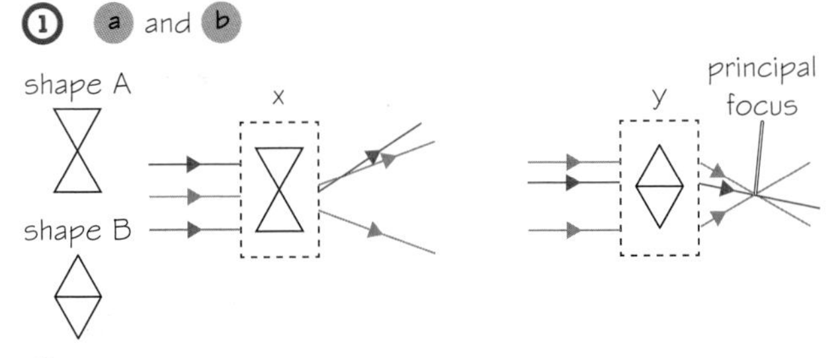

2 Name Shape Symbol

Name	Shape	Symbol
convex		
concave		

3

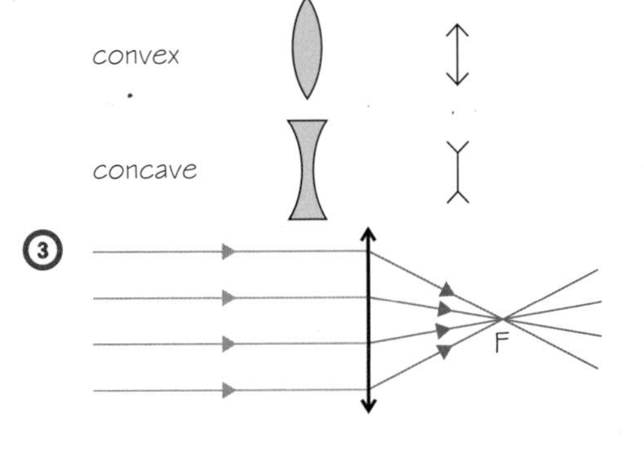

4 a and b

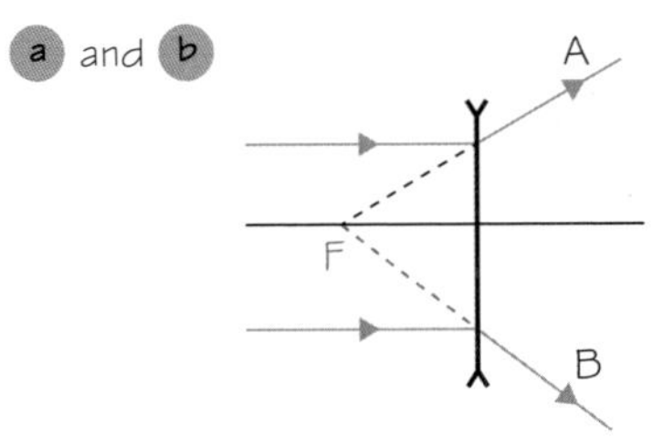

Page 13

1 a, b and c

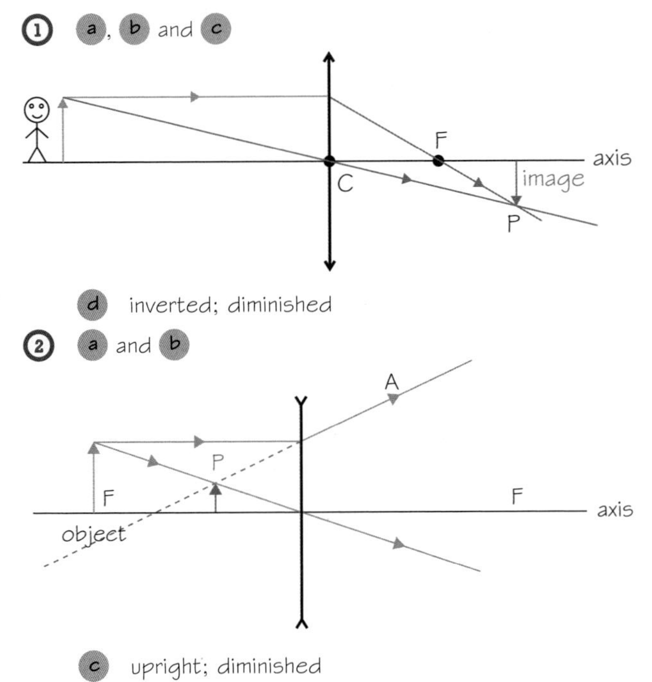

d inverted; diminished

2 a and b

c upright; diminished

Page 14

1 a No b convex c arrows

2 Light ray on right passing through F should be labelled A.

3 The arrow which forms the image has been drawn incorrectly. The head of the arrow is in the right place (where the rays meet) but the arrow points in the wrong direction. Always draw the arrowhead pointing away from the axis with the bottom of the arrow on the axis. In this question, the arrow should be upside down (the image is inverted).

Page 15

1 convex

2 a no

b 'Dot them back' to find where they meet.

3 a

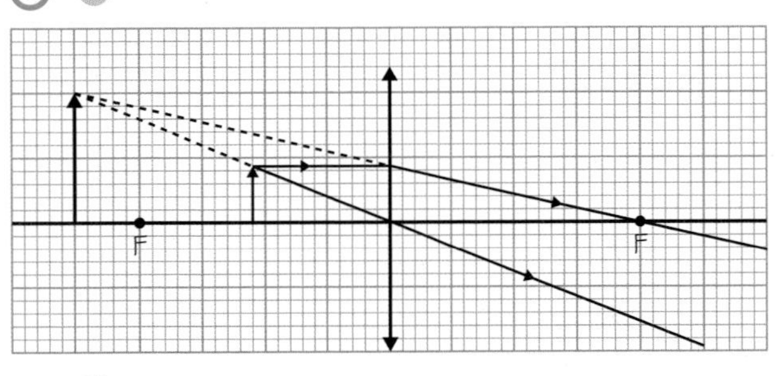

b upright; magnified; virtual

Page 16

Exam-style question

1

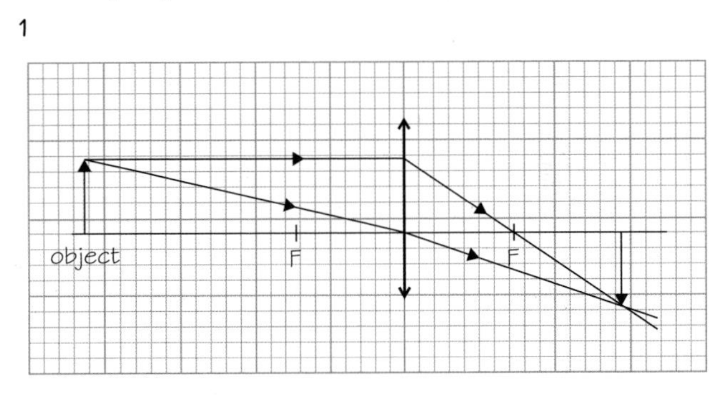

Unit 3

Page 18

1 a violet, indigo, blue, green, yellow, orange, red

b If light is red-shifted, it moves towards the red end of the spectrum, so the wavelength **increases** and the frequency **decreases**.

2 B sound waves from a car radio driving away from you

C two bicycles racing around a track at 13 m/s and 14 m/s

F two marbles rolling at the same speed in different directions

③

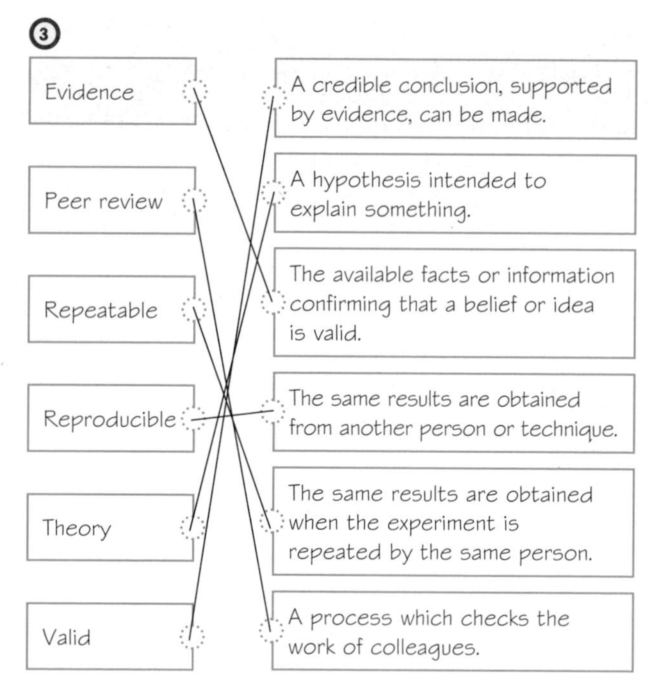

Page 19

① The waves in front of the duck have a shorter wavelength and higher frequency. The waves behind the duck have a longer wavelength and lower frequency.

② When a source moves towards an observer, the observed wavelength **decreases** and the frequency **increases**.

When a source moves **away from** an observer, the observed wavelength increases and the frequency **decreases**.

③

light source	wave pattern	Spectrum
stationary		no shift
moving towards you		blue-shift
moving away from you		red-shift

increasing wavelength/decreasing frequency

Page 20

Evidence	Interpretation of the evidence
Light coming to us from most other galaxies is red-shifted.	Those galaxies are moving **away from** us.
The further away a galaxy is, the **more** the light is red-shifted.	The further away a galaxy is, the **faster** it is moving.

b bigger

② Big Bang

③ Steady State

④ They both suggest that the universe is expanding.

⑤ both

⑥ Big Bang

Page 21

① red-shift

② a Big Bang b Steady State

③ The Big Bang theory is supported by **red-shift**, which suggests that the universe is **expanding**. However, the **Steady State** theory is also supported by this observation. Other evidence has been found which **only** supports the **Big Bang** theory, so this is the currently **accepted** model. Scientists continue to investigate in order to develop and improve established models.

④ a There is still a lot about the universe that is not understood, for example **dark** energy and dark **mass**.

b Since 1998, observations of supernovae suggest that distant galaxies are receding ever faster.

This means that the rate of expansion of the universe is **accelerating**.

Scientists think that dark energy may explain this. It is not well understood as it cannot be seen and only its effects can be observed.

c Dark mass may explain the rate of rotation of galaxies. Scientists gain a lot of information by observing **radiation** coming from space. Dark mass is not well understood as it cannot be directly **observed** because it is **dark**.

Page 22

① a no b no

c It is when light from a galaxy moving away from you is shifted to the red end of the spectrum (1), meaning its wavelength increases and its frequency decreases (1).

② a Explain b description c no

d

What the red-shift of light from distant galaxies tells us	True	False
The universe is expanding	✓	
All galaxies move at the same speed		✓
The closer a galaxy is to the observer, the faster it moves		✓
The faster the galaxy moves, the bigger the increase in wavelength	✓	
Red-shift is larger for galaxies that are further away	✓	
The further away a galaxy is from the observer, the faster it moves	✓	

e The red-shift of light is larger for galaxies that are further away **(1)** so the further away a galaxy is from us, the faster it is moving. **(1)** This tells us that galaxies are moving away from us and the universe is expanding. **(1)**

Page 23

1.1 (i) least

(ii) Galaxy A **(1)** has the smallest red shift so galaxy A **(1)** has the slowest speed. The galaxy with the slowest speed will be the nearest, so galaxy A **(1)** is nearest to us.

1.2 The Big Bang theory is supported by the evidence of red-shift because it predicts that the universe is expanding **(1)**. However, other observations (such as cosmic microwave background radiation coming from all over the sky) are only supported by the Big Bang theory (because huge amounts of radiation were predicted to have been released at the start of the universe) **(1)** so it currently has the most evidence to support it.

1.3 dark mass / dark energy **(1)**

Page 24

1 The Big Bang theory suggests that the universe began from a single point with a huge explosion which released immense amounts of radiation. The Steady State theory suggests that the universe has always been here and it continuously creates new matter as it expands **(1)**.

Both the Big Bang theory and the Steady State theory suggest that the universe is expanding **(1)**. Evidence from the red-shift of light from distant galaxies shows us that the galaxies are moving away from us, so this evidence supports both theories **(1)**.

However, other new evidence and observations have been made that can only be explained by the Big Bang theory and not the Steady State theory **(1)**. Therefore, the Big Bang is currently the most supported theory.

Unit 4

Page 26

1

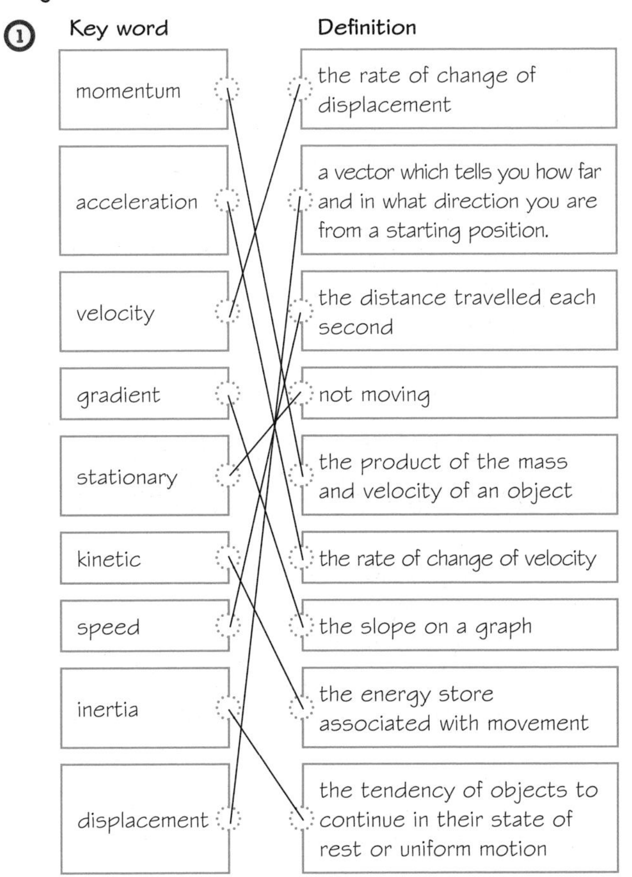

2

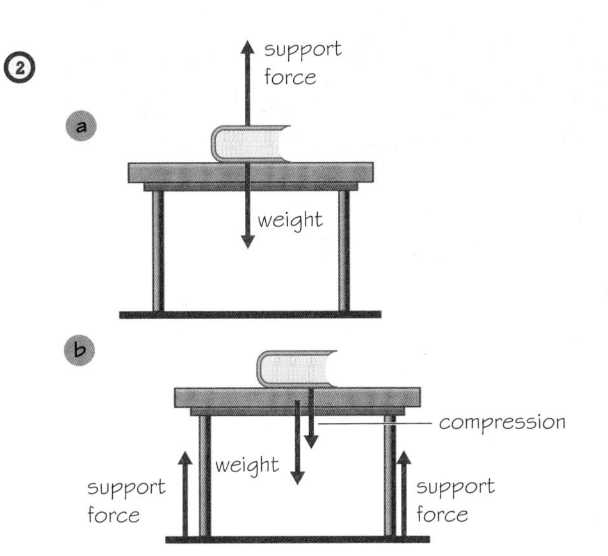

Page 27

1 a stationary; straight; speed

b proportional; inversely; mass

c equal; opposite

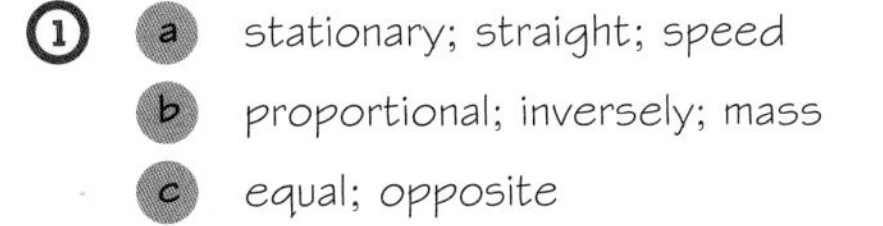

2

Quantity	Symbol	Unit
force	a	newton, N
mass	F	metre per second squared, m/s²
acceleration	m	kilogram, kg

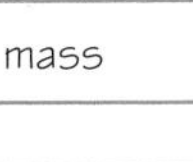

3

Resultant force	↓ 6.0 N	→ 2.0 N	↓ 120 N
Acceleration	$a = \frac{F}{m} = \frac{6.0}{3.0}$ $a = 2.0$ m/s²	$a = \frac{F}{m} = \frac{2.0}{0.2}$ $a = 10$ m/s²	$a = \frac{F}{m} = \frac{120}{6.0}$ $a = 20$ m/s²

Page 28

1 first row: 4 m/s

second row, second column: −3

second row, third column: −9; −12 m/s

2

Calculation 1	Calculation 2
A car has a mass of 1200 kg. Calculate the force needed to accelerate it from 0 m/s to 5.0 m/s in 5.0 s.	A lorry has a mass of 8000 kg. Calculate the force needed to accelerate it from 9.0 m/s to 1.0 m/s in 9.0 s.
0 to 5.0 m/s $\Delta v = 5.0 - 0 = 5.0$ m/s	9.0 to 1.0 m/s $\Delta v = 1 - 9 = -8.0$ m/s
$a = \frac{\Delta v}{t} = \frac{5.0}{5.0}$ $a = 1.0$ m/s	$a = \frac{\Delta v}{t} = \frac{-8.0}{9.0}$ $a = -0.888888$ m/s
$F = m\,a$ $F = 1200 \times 1.0$ $F = 1200$ N	$F = m\,a$ $F = 8000 \times -0.888888$ $F = -7111$ N (to 4 significant figures)

Page 29

1. **a** A B D E **b** A D **c** A D
2. Column A: $p = m\,v$; $p = 5000 \times 8.5$; 42 500 (or 4.25×10^4) kg m/s to the right

   Column B: $p = m\,v$; $p = 42\,500$ kg $\times$ 7.5; 318 750 (or 3.19×10^5 kg m/s) to the left
3. Column A: $4.0 \times 1.5 = 6.0$ to the right; $3.0 \times 2.0 = 6.0$ to the right; $6.0 + 6.0 = +12.0$ kg m/s to the right

   Column B: $p = m\,v$; $8.0 \times 3.0 = 24.0$ to the right; $6.0 \times 4.0 = 24.0$ to the left; $24.0 - 24.0 = 0$ kg m/s

Page 30

1. **a** $m = \frac{p}{v}$ **b** $\frac{0.3}{2.5} = 0.12$ kg

   c The student has used far too many significant figures. The data in the question has only two s.f. and so the answer should have the same number.

   d The student has used the correct form of the equation ($p = m\,v$).

   e The student has not found the change in velocity. They have subtracted the numbers but not taken into account the direction. The real change in velocity is 4.50 m/s. The student would get marks even though they used the wrong mass as they did not score the mark for the calculation earlier.

   f change in momentum = change in velocity × mass = $4.50 \times 0.12 = 0.54$ kg m/s

Page 31

Exam-style question

1.1 0

1.2 $p = m\,v = 0.045 \times 80 = 3.6$ **(1)** kg m/s **(1)**

1.3 [–]3.6 kg m/s **(1)** allow error carried forward from part 1.2

1.4 $a = \frac{\Delta v}{t} = \frac{(80-0)}{0.02} = 4000$ **(1)** m/s² **(1)**

1.5 $F = m\,a = 0.045 \times 4000 = 180$ N **(1)** (allow error carried forward from 1.4)

Page 32

Exam-style question

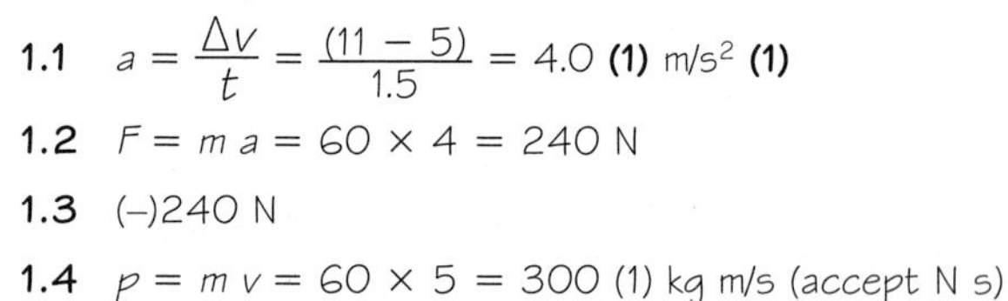

1.1 $a = \frac{\Delta v}{t} = \frac{(11 - 5)}{1.5} = 4.0$ **(1)** m/s² **(1)**

1.2 $F = m\,a = 60 \times 4 = 240$ N

1.3 (–)240 N

1.4 $p = m\,v = 60 \times 5 = 300$ (1) kg m/s (accept N s)

Unit 5

Page 34

1. **a**, **b**, **c**

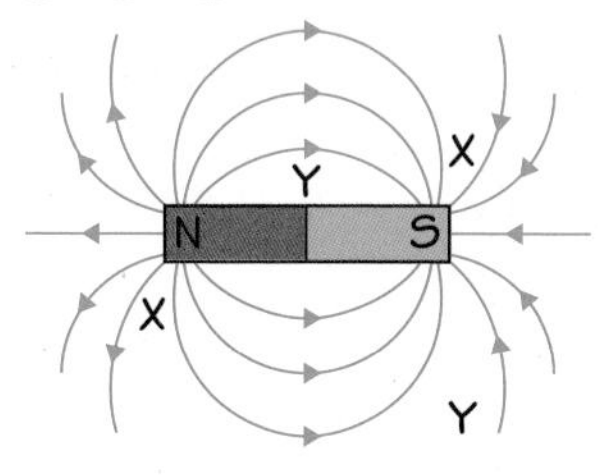

Other places for X and Y can be chosen (for example one of the Y locations shown in the diagram). X should be where the magnetic field lines are close together and Y should be where they are far apart.

2.

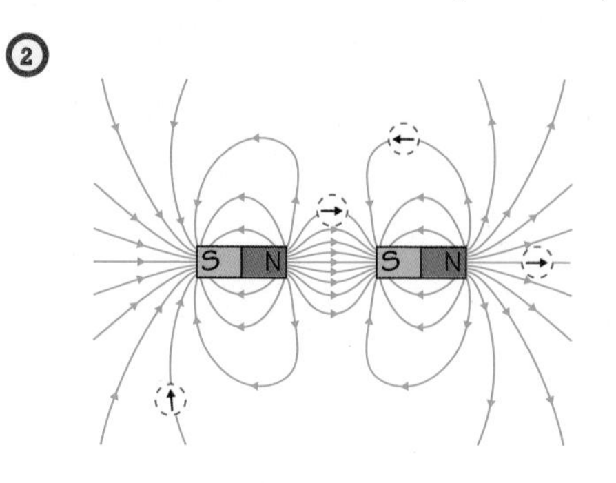

3. strong; uniform

Page 35

1. current; magnetic field
2. The current is moving upwards.
3. strongest; weaker; stronger
4. True; False; True

Page 36

1. **a**, **b**, **c**

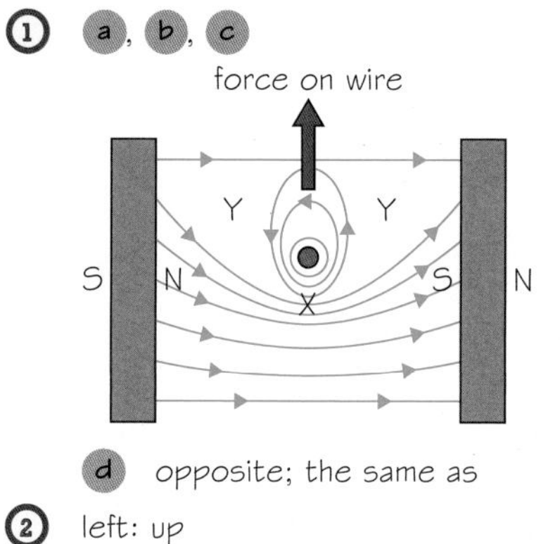

   d opposite; the same as

2. left: up

   middle: left

   right: no force

Page 37

1.

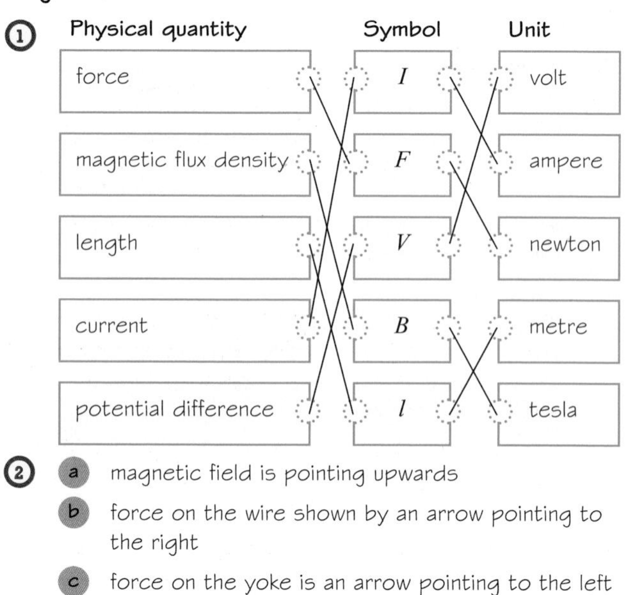

2. **a** magnetic field is pointing upwards

   b force on the wire shown by an arrow pointing to the right

   c force on the yoke is an arrow pointing to the left

(d) From $F = B\ I\ l$, double the current, double the strength of the magnetic field or double the length of the wire in the magnetic field.

(3) $F = B\ I\ l$

$F = 0.2 \times 1.5 \times 0.12$

$F = 0.036$ N

Page 38

(1) (a)

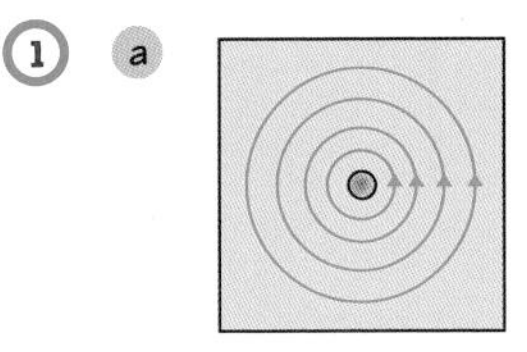

(b) The spacing between circles increases with distance from the wire because the magnetic field strength decreases with distance from the wire. The field lines need arrows to show the direction of the magnetic field. In this case, the field is anticlockwise as predicted by the right-hand grip rule.

(2) (a) The letter X would have been better in the middle of the solenoid where the field lines are closest together and not diverging.

(b) Other places of weaker magnetic field strength are shown on the diagram:

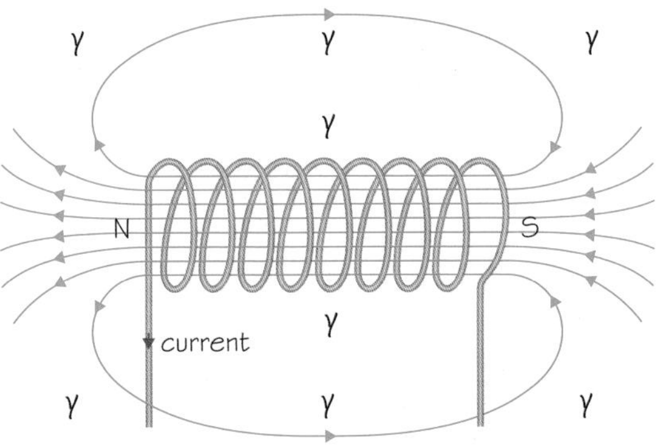

(c) The magnetic field is **strong** because the lines are close together and **uniform** because the lines are straight, parallel and evenly spaced.

Page 39

Exam-style question

1.1 A force acts on the wire because the current in the wire and the magnetic field from the magnets are at right-angles to each other **(1)**. The magnetic field around the wire interacts with the magnetic field between the magnets **(1)** to create a force between the wire and magnets **(1)**.

1.2 There should be an upward arrow on the wire between X and Y. (The battery shows that the current is moving around the circuit in a clockwise direction. Fleming's left-hand rule allows you to predict the direction of the force on the wire.)

1.3 force on a conductor (at right-angles to a magnetic field) carrying a current = magnetic flux density × current × length ($F = B\ I\ l$)

1.4 Increase the magnetic flux density (allow magnetic field strength) **(1)**; increase the current **(1)**; increase the length of wire in the magnetic field **(1)**

Page 40

Exam-style question

1.1 $F = m\ g = 0.016 \times 10$ **(1)** $= 0.16$ **(1)** N **(1)**

1.2 $F = B\ I\ l$ **(1)**

$B = \frac{F}{I\ l} \times \frac{0.16}{3.2 \times 0.1} = \frac{0.16}{0.32}$ **(1)**

$B = 0.5$ **(1)** T **(1)** (allow errors carried forward from 1.1)

Unit 6

Page 42

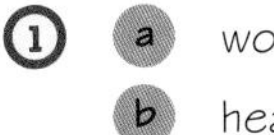

(1) (a) work done by a force against friction

(b) heating

(c) work done by an electric current

(2) an electric current; a force; heating

(3) Store that decreases (from top): gravitational store of the object; elastic store of the bow and bow string;

Stores that increase (from top): kinetic energy of particles increasing speed; thermal store of the water (and surroundings); kinetic store of the arrow.

(4) (a) 500 (b) 2.5 (c) 550

Page 43

(1)

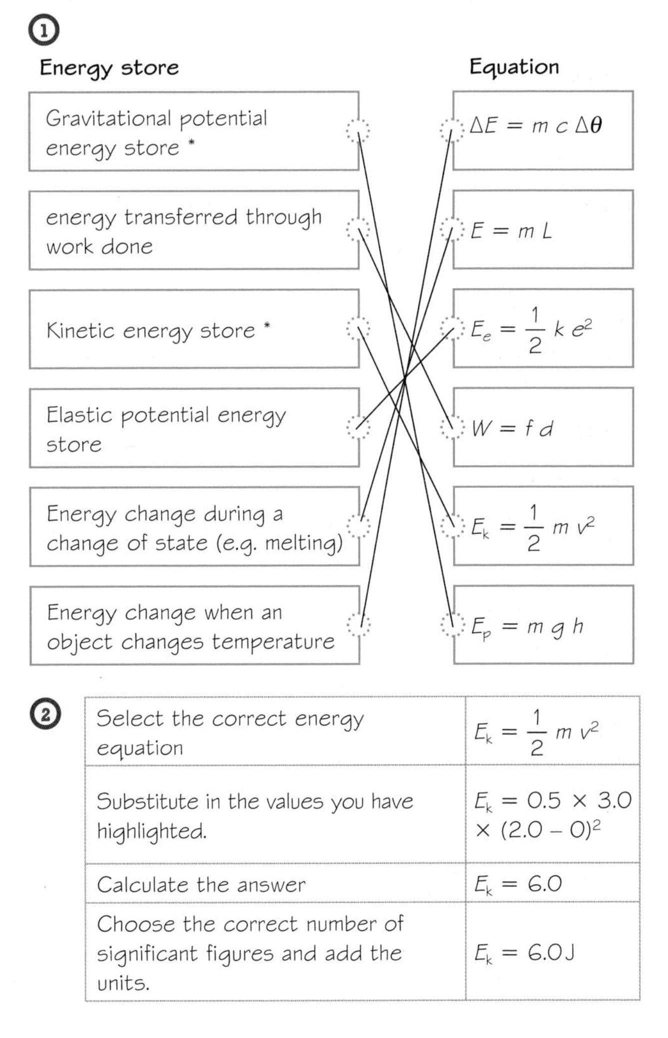

(2)

Select the correct energy equation	$E_k = \frac{1}{2} m\ v^2$
Substitute in the values you have highlighted.	$E_k = 0.5 \times 3.0 \times (2.0 - 0)^2$
Calculate the answer	$E_k = 6.0$
Choose the correct number of significant figures and add the units.	$E_k = 6.0$ J

Page 44

1 a The spring used to launch the darts has a spring constant of 200 N/m.

When a dart is launched the spring is compressed by 0.50 m. The efficiency of the spring launcher is 0.90.

b $E_e = \frac{1}{2} ke^2$

c $E_e = \frac{1}{2} ke^2 = 0.5 \times 200 \times 0.50^2 = 25\text{ J}$

d **Method 1**

$$\text{efficiency} = \frac{\text{useful output energy transfer}}{\text{total input energy transfer}}$$

$$0.9 = \frac{\text{useful output energy transfer}}{25}$$

$$25 \times 0.9 = \frac{\text{useful output energy transfer}}{\cancel{25}} \times \cancel{25}$$

22.5 J = useful output energy transfer

Method 2

$$\text{efficiency} = \frac{\text{useful output energy transfer}}{\text{total input energy transfer}}$$

useful energy transferred = total energy input × efficiency

= 25 × 0.90 = 23 J (22.5 J)

Page 45

1 The number of collisions increases. The pressure will increase.

2 The volume will be halved to 30 cm^3.

3 Before: $p\,V = 850\,000 \times 135$

$p\,V$ = constant

After: final pressure × 45 = 850 000 × 135

$$\text{final pressure} = \frac{(850\,000 \times 135)}{45} = 2\,550\,000\text{ Pa}$$

4 increases; increases; increases

5 a $P = \frac{f}{a} = \frac{12}{0.3} = 40\text{ Pa}$

b $W = f\,d = 12 \times 1.2 = 14.4\text{ J}$

Page 46

1 a Gravitational potential energy $E_p = m\,g\,h = 1.0 \times 10 \times 3.6 = 36\text{ J}$

b The student has written v instead of v^2.

$$v = \sqrt[2]{\frac{2E_k}{m}} = \sqrt[2]{\frac{2 \times 36}{1.0}} = 8.5\text{ ms}^{-1}$$

2 a 27 000 J

b $\frac{\Delta E}{m\,c}$

c °C

d $\Delta\theta = \frac{\Delta E}{m\,c} = \frac{27\,000}{(0.5 \times 4200)} = 12.9\text{°C}$

Page 47

Exam-style question

1.1 There is a transfer from the mechanical **(1)** store of the bicycle pump to the kinetic store of the particles (kinetic energy) **(1)**.

1.2 $\text{pressure} = \frac{\text{force}}{\text{area}}$ and work = force × distance **(1)**

work done (J) = pressure (Pa) × area (m^2) × distance moved (m) **(1)**

= 1.2 × 0.125 × 0.3 = 0.045 J **(1)**

1.3 The average **speed** of the gas particles **increases**, which is detected as a rise in temperature **(1)** (Both words are required for the mark.)

1.4 Before: $p\,V = 1.4 \times 10^5 \times 25$

$p\,V$ = constant

After: final pressure × 7 = 1.4 × 10^5 × 25

$$\text{final pressure} = \frac{(1.4 \times 10^5 \times 25)}{7}\ \textbf{(1)}$$

$$= \frac{3\,500\,000}{7}$$

= 500 000 Pa or 5 × 10^5 Pa **(1)**

Page 48

Exam-style question

1 1.1 $E_p = m\,g\,h$ **(1)** = 0.25 × 10 × 2.0 = 5.0 J **(1)**

1.2 $E_k = \frac{1}{2} m\,v^2$ **(1)**, $v = \sqrt[2]{\frac{2E_k}{m}}$ **(1)** = 6.3 m/s **(1)**

(Allow error carried forwards from 1.1.)

1.3 As the ball hits the ground it deforms so kinetic → elastic **(1)**; as the ball rebounds elastic → kinetic **(1)**.

1.4 Calculation of gravitational potential energy after bounce (3.5 J) **(1)**, use of

$$\text{efficiency} = \frac{\text{useful output energy transfer}}{\text{total input energy transfer}}$$ **(1)** to give efficiency of 0.70 **(1)**. (Allow use of ratio of heights to reach efficiency.)

1.5 As the tennis ball warms up, the pressure increases **(1)** because the volume of the ball is fixed. The work done on the ball increases the kinetic energy **(1)** of the particles. This causes an increase in the temperature of the gas **(1)**.

Unit 7

Page 50

1 first row: positive; in the nucleus

second row: neutral; in the nucleus

third row: negative; different energy levels around the nucleus

2 a protons

b protons; neutrons

c mass number; atomic number

d electrons; nucleus

3 a All of the chlorine isotopes have the same number of protons (17).

b All of the chlorine isotopes have different numbers of neutrons (18, 19 and 20).

4

Isotope and chemical symbol	Protons	Neutrons	Electrons	Atomic Notation
Carbon-14 C	(from atomic number) 6	Mass no. – atomic no. 14 – 6 = 8 8	(same as protons) 6	$^{14}_{6}C$
Carbon-12 C	6	6	6	$^{12}_{6}C$

Page 51

1 a protons b 2 c 4
d $^{4}_{2}He$

2 $^{218}_{86}Rn$; $^{6}_{3}Li$

3 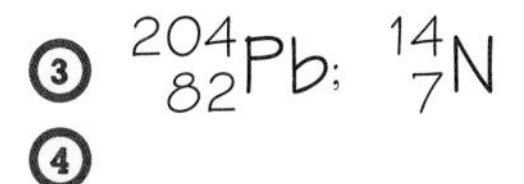
$^{204}_{82}Pb$; $^{14}_{7}N$

4

Decay type	Equation	Decay type	Equation
alpha	$^{185}_{79}Au \rightarrow$ $^{181}_{77}Ir + ^{4}_{2}He$	alpha	$^{231}_{91}Pa \rightarrow$ $^{227}_{89}Ac + ^{4}_{2}He$
beta	$^{14}_{6}C \rightarrow$ $^{14}_{7}N + ^{0}_{-1}e$	beta	$^{8}_{3}Li \rightarrow$ $^{8}_{4}Be + ^{0}_{-1}e$

Page 52

1

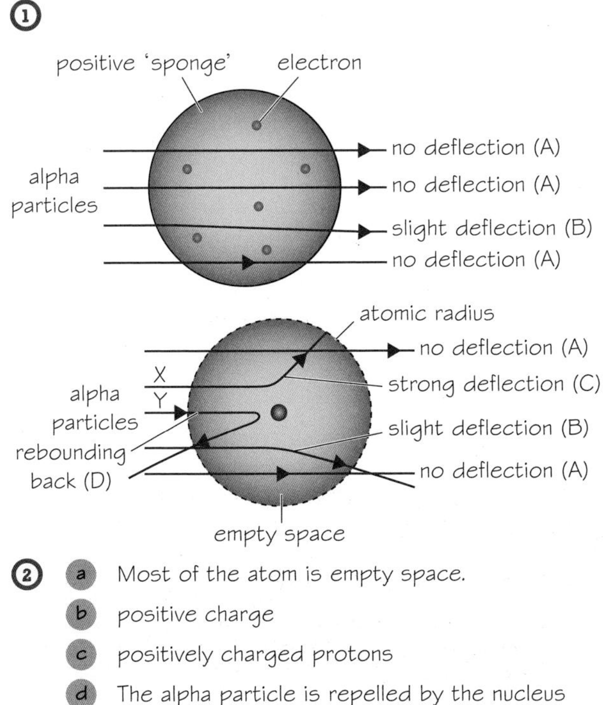

2 a Most of the atom is empty space.
b positive charge
c positively charged protons
d The alpha particle is repelled by the nucleus because they have the same type of electric charge.
e The alpha particle is repelled much more strongly because it passes closer to the nucleus and the force is much greater.

Page 53

1 a 1800 Bq b 5 hours c 5 hours

2 $\frac{1}{16}$

3

Time in hours	0	→ 1st half-life	5	→ 2nd half-life	10	→ 3rd half-life	15
Activity in Bq	1800		900		450		225
Fraction remaining	$\frac{1}{1}$		$\frac{1}{2}$		$\frac{1}{4}$		$\frac{1}{8}$

4

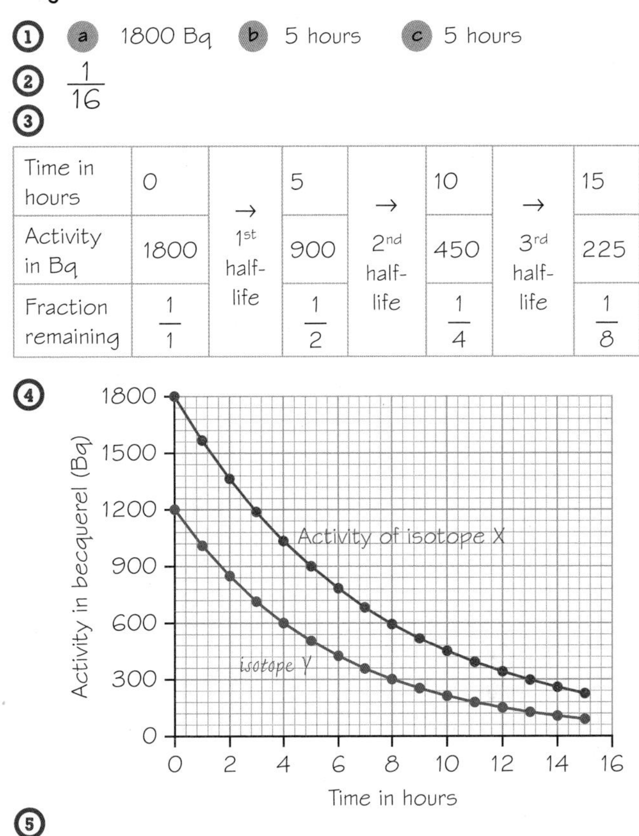

5

Time in hours	0	→ 1st half-life	3.5	→ 2nd half-life	7	→ 3rd half-life	10.5
Activity in Bq	1200		600		300		150
Fraction remaining	$\frac{1}{1}$		$\frac{1}{2}$		$\frac{1}{4}$		$\frac{1}{8}$

Page 54

1 It means one half-life has passed.

2 a Four. There are four arrows.
b 16 000 Bq → 8000 Bq → 4000 Bq → 2000 Bq → 1000 (the answer is 1000 Bq)

3 a 90; 0; e
b $^{90}_{38}Sr \rightarrow ^{90}_{39}Y + ^{0}_{-1}e$

Page 55

Exam-style question

1.1 Atoms of the same element (atoms with the same number of protons) with different mass numbers / numbers of neutrons.

1.2 source A

1.3 source B

1.4 4.5 days

1.5 $^{131}_{53}I \rightarrow ^{131}_{54}Xe + ^{0}_{-1}e$

Page 56

Exam-style question

1.1

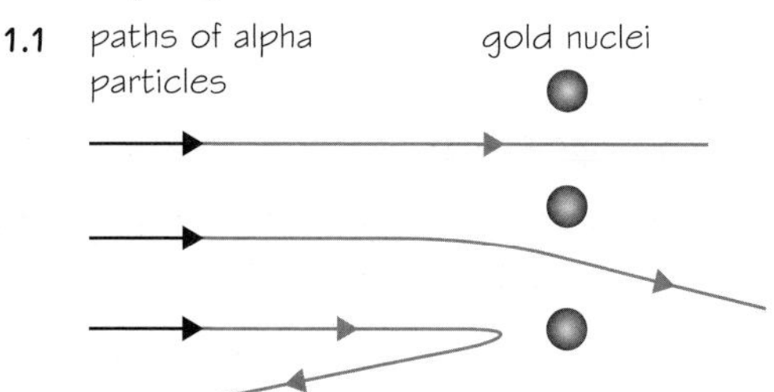

1.2 The alpha particles and nuclei both have positive charge **(1)** and so repel each other. **(1)**

Unit 8

Page 58

① pressure; power

volume; potential difference; volt

②

Distance or length metre	Time second	Energy joule
Power watt	Mass kilogram	Current ampere
Force newton	Potential difference volt	Pressure pascal

③ kg m/s (the unit of mass multiplied by the unit of velocity)

④ $a = \frac{F}{m}$

Page 59

① (a) current of 1.6 A; 25 s (b) charge

(c) I, Q, t (d) $Q = I\,t$

(e) $Q = I\,t$

$Q = 1.6 \times 25$

$Q = 40\,C$

② $P = I^2 R = 8^2 \times 3 = 192\,W$

Page 60

① (a) volume = 0.1 × 0.1 × 0 1 = 0.001 m^3

(b) $D = \frac{m}{V} = \frac{2.7}{0.001} = 2700\,kg/m^3$

②

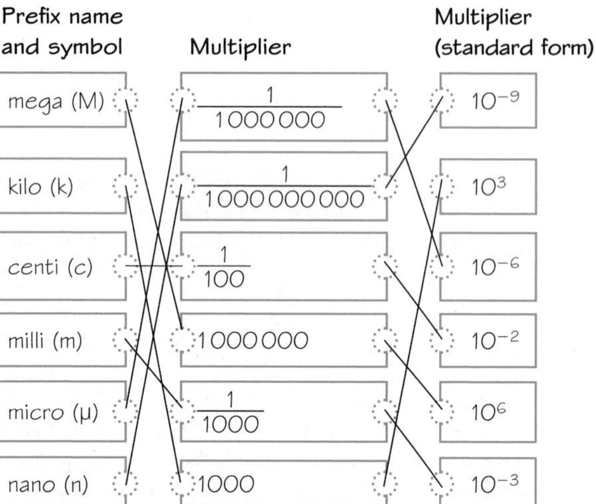

③

145 m = 0.145 km	145 m = 14 500 cm	2440 mm = 2.44 m
97.7 MHz = 97 700 000 Hz	48 mV = 0.048 V	101 300 Pa = 101.3 kPa
2300 W = 2.3 kW		

④

Object	Distance	Time	Average speed
car	240 m	40 s	6 m/s
lizard	300 cm	15 s	20 cm/s
rocket	480 km	64 s	7.5 km/s

Page 61

① 2; 5; 1; 6; 4; 3

② $\Delta\theta = 16\,°C$; $c = \frac{\Delta E}{m\,\Delta\theta}$; $c = 391\,J/kg\,°C$

Page 62

① (a) Did not convert the time in minutes to seconds. Gave the answer without a unit.

(b) $E = P\,t$

$E = 50 \times (3 \times 60)$

$E = 9000\,J = \Delta E$

(c) Although the student used the incorrect value for ΔE, they chose the correct equation, rearranged it correctly, and calculated the answer correctly.

(d) $\Delta E = m\,c\,\Delta\theta$

$c = \frac{\Delta E}{m\,\Delta\theta}$

$c = \frac{9000}{(0.5 \times 45)}$

$c = 400\,J/kg\,°C$

Page 63

Exam-style question

1.1 $P = V\,I$ **(1)** $= 8 \times 4 = 32$ W **(1)**

1.2 $E = P\,t = 32 \times 300 = 9600\,J$

1.3 $E = m\,L$

$L = \frac{E}{M} = \frac{9600}{0.0343} = 280\,000\,J/kg$

Page 64

Exam-style question

1.1 $P = I^2\,R$

1.2 $P = 5000^2 \times 2.7 = 25\,000\,000 \times 2.7 = 67\,500\,000\,W = 67.5\,MW$

1.3 $P = 2500^2 \times 2.7 = 6\,250\,000 \times 2.7 = 16\,875\,000 = 16.9\,MW$

The power loss in the second power line is one-quarter of the loss in the first power line.

Unit 9

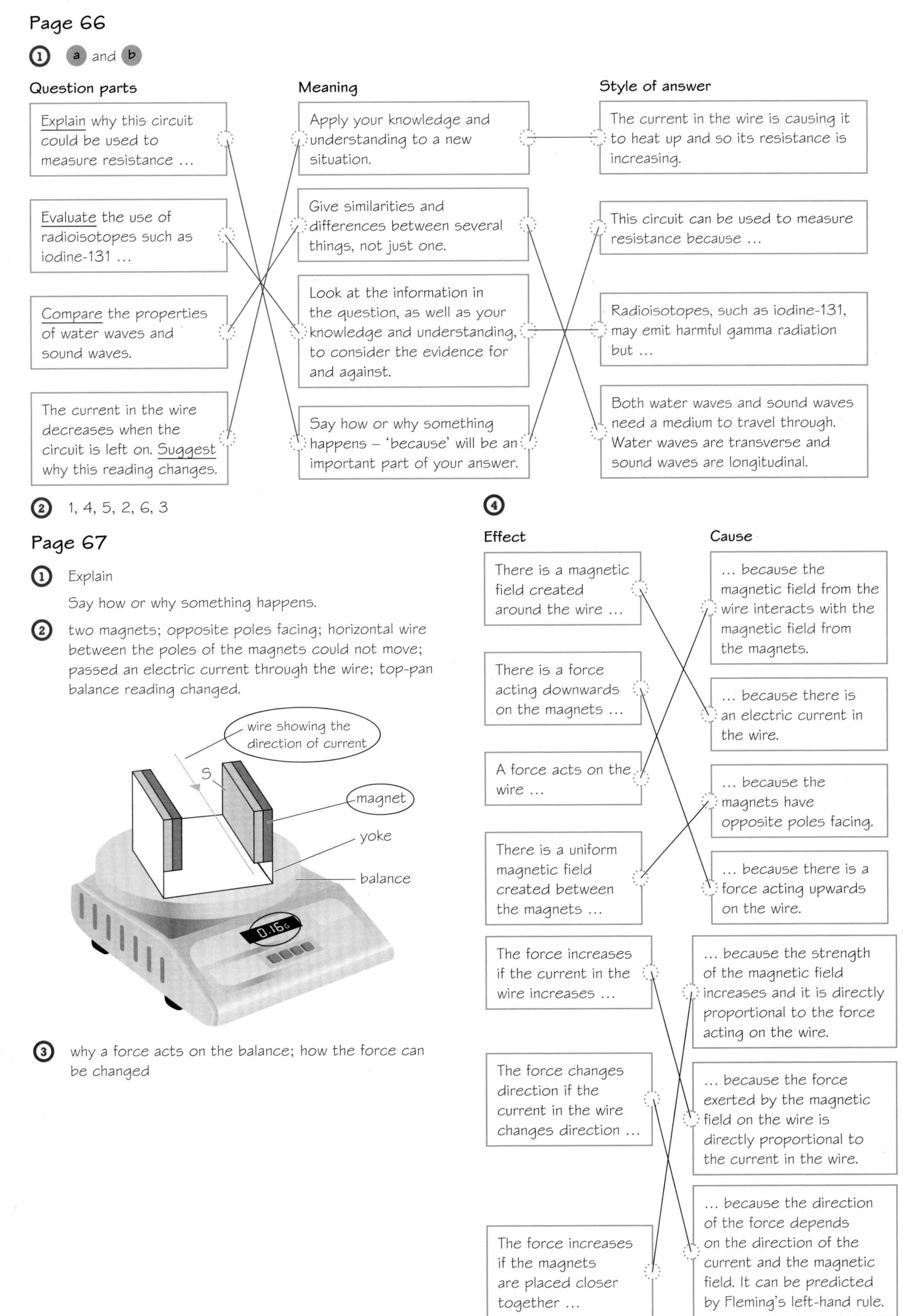

Page 66
1 a and b
Question parts
Explain why this circuit could be used to measure resistance …
Evaluate the use of radioisotopes such as iodine-131 …
Compare the properties of water waves and sound waves.
The current in the wire decreases when the circuit is left on. Suggest why this reading changes.
Meaning
Apply your knowledge and understanding to a new situation.
Give similarities and differences between several things, not just one.
Look at the information in the question, as well as your knowledge and understanding, to consider the evidence for and against.
Say how or why something happens – 'because' will be an important part of your answer.
Style of answer
The current in the wire is causing it to heat up and so its resistance is increasing.
This circuit can be used to measure resistance because …
Radioisotopes, such as iodine-131, may emit harmful gamma radiation but …
Both water waves and sound waves need a medium to travel through. Water waves are transverse and sound waves are longitudinal.
2 1, 4, 5, 2, 6, 3
Page 67
1 Explain
Say how or why something happens.
2 two magnets; opposite poles facing; horizontal wire between the poles of the magnets could not move; passed an electric current through the wire; top-pan balance reading changed.
wire showing the direction of current
S
magnet
yoke
balance
0.16
3 why a force acts on the balance; how the force can be changed
4
Effect
There is a magnetic field created around the wire …
There is a force acting downwards on the magnets …
A force acts on the wire …
There is a uniform magnetic field created between the magnets …
The force increases if the current in the wire increases …
The force changes direction if the current in the wire changes direction …
The force increases if the magnets are placed closer together …
Cause
… because the magnetic field from the wire interacts with the magnetic field from the magnets.
… because there is an electric current in the wire.
… because the magnets have opposite poles facing.
… because there is a force acting upwards on the wire.
… because the strength of the magnetic field increases and it is directly proportional to the force acting on the wire.
… because the force exerted by the magnetic field on the wire is directly proportional to the current in the wire.
… because the direction of the force depends on the direction of the current and the magnetic field. It can be predicted by Fleming's left-hand rule.

Page 68

1. The motor effect
2. 11, 8, 7, 5, 2, 4, 1, 6, 10, 9, 3
3. Sample answer: There is a magnetic field created around the wire because there is an electric current in the wire.

   There is a uniform magnetic field created between the magnets because the magnets have opposite poles facing. A force acts on the wire because the magnetic field from the wire interacts with the magnetic field from the magnets.

   There is a force acting downwards on the magnets because there is a force acting upwards on the wire from Fleming's left-hand rule.

   The force increases if the current in the wire increases because the force exerted by the magnetic field on the wire is directly proportional to the current in the wire.

   The force increases if the magnets are placed closer together because the strength of the magnetic field increases and the force acting on the wire is directly proportional to the strength of the magnetic field.

   The force changes direction if the current in the wire changes direction because the direction of the force depends on the direction of the current and the magnetic field. It can be predicted by Fleming's left-hand rule.

Page 69

1. a air–glass boundary; inside the glass block; glass–air boundary

   b reflection; refraction; absorption; transmission
2. Reflection and refraction; Absorption and emission of wave energy; The law of conservation of energy
3. a

Region	Reflection	Refraction	Absorption	Transmission
air–glass boundary	some light is reflected	most light is refracted		
inside the glass block			some light is absorbed	most light is transmitted
glass–air boundary	some light is reflected	most light is refracted		

   b some light is absorbed but the remaining light energy is transmitted by the block because all energy from the light ray inside the block must be absorbed or transmitted;

   some light is reflected internally but the remaining light is refracted at the surface and leaves the block because all energy from the light ray must be refracted or reflected.

Page 70

1. Copper is a good conductor but is not normally magnetic; If you make the current in the rod larger then the reading on the balance will change more because there will be an even larger force because of the equation $F = B I l$.

2. a This magnetic rod is then affecting the magnets because there is a force whenever two magnets are placed near to each other because of their magnetic fields.

   The magnets act to repel each other so the rod is pushed away from the permanent magnets.

   b When magnets put a force on the rod it is the same force that pushes back on the magnets and then the top-pan balance.
3. Sample answer: The current in the rod creates a circular magnetic field around it. This field interacts with the field from the magnets to produce a pair of forces acting in opposite directions on the rod and the magnets. The direction of the force from the frame on the rod is determined using Fleming's left-hand rule and is downwards. The rod is prevented from moving as it is clamped, so it exerts an equal and opposite upward force on the frame holding the magnets. There is a reduced resultant force on the balance from the frame, which decreases the balance reading.

Page 71

1. Explain
2. very high potential differences are used to transfer electrical power in overhead power cables; lower potential differences are used in houses
3. To transmit a large power, the current must be high or the voltage must be high (or both).
4. Use low resistances or low currents (or both).
5. A high voltage will allow a large power without a large heating effect in the overhead cable.
6. Explain power losses in cables using $P = I^2 R$.

   Need for transmission at high voltage.
7. Answer could include the following points in a logical order for 6 marks:

   The overhead power cables waste power because of electrical heating. To reduce this waste, a high voltage can be used because this allows the same power to be transmitted with a much lower current (from $P = V I$). This lower current means there is a lot less power wasted due to electrical heating in the cable (from $P = I^2 R$).

Page 72

Exam-style question

1. Answer could include the following points in a logical order for 6 marks:

   Both electromagnetic waves and sound waves transfer energy from the wave source to the surroundings. Both can be reflected or absorbed by a surface. Both electromagnetic waves and sound waves can be refracted at a boundary if they change speed. They both have different wavelengths and frequencies, which are related by wave speed = frequency × wavelength.

   Electromagnetic waves can travel through a vacuum, but sound waves need a material to travel through. Electromagnetic waves are transverse waves, but sound waves are longitudinal waves. (Electromagnetic waves are very fast, but sound waves are about a million times slower. Some electromagnetic waves can cause ionisation, but sound waves cannot.)